ECOSYSTEMS

access to geography

ECOSYSTEMS

Meg Gillett

Hodder Murray

A MEMBER OF THE HODDER HEADLINE GROUP

For: Alistair, Calum, Greg, Harriet, Hector, Helen, Jack, Jonathan, Kris, Loredana, Lucie, Luke, Matthew, Nick, Nik, Richard, Sam M, Sam P, Steph, Tom F, Tom H and Tom R.

The publishers would like to thank the following individuals, institutions and companies for permission to reproduce copyright illustrations in this book:
Michael Fogden/Oxford Scientific, page 83; South West News Service, swns.com, page 67; © Roger Tidman/CORBIS, page 120.

Every effort has been made to trace all copyright holders, but if any have been inadvertently overlooked the Publishers will be pleased to make the necessary arrangements at the first opportunity.

Although every effort has been made to ensure that website addresses are correct at time of going to press, Hodder Murray cannot be held responsible for the content of any website mentioned in this book. It is sometimes possible to find a relocated web page by typing in the address of the home page for a website in the URL window of your browser.

Orders: please contact Bookpoint Ltd, 130 Milton Park, Abingdon, Oxon OX14 4SB. Telephone: (44) 01235 827720. Fax: (44) 01235 400454. Lines are open 9.00–6.00, Monday to Saturday, with a 24-hour message answering service. Visit our website at www.hoddereducation.co.uk

First published in 2005 by
Hodder Murray, an imprint of Hodder Education,
a member of the Hodder Headline Group
338 Euston Road
London NW1 3BH

Impression number 10 9 8 7 6 5 4 3 2 1
Year 2010 2009 2008 2007 2006 2005

Cover photo of the woodlands near Vaduz Castle in Lichtenstein by Michael Hill
Typeset in 10/11pt Baskerville and produced by Gray Publishing, Tunbridge Wells
Printed in Malta

A catalogue record for this title is available from the British Library

ISBN-10: 0340 88920 9
ISBN-13: 978 0340 88920 6

Contents

1 Systems and Ecosystems

KEY WORDS

Abiotic environment non-living environment comprising energy and matter
Adaptation the way in which an organism changes in response to abiotic factors within its habitat in order to be able to survive
Biodiversity the number and variety of living organisms within a habitat
Biomass total mass of vegetation within a specific area
Biosphere (also called **ecosphere**) zone of earth in which all life occurs
Climax (mature) community final stage of vegetation succession, reached when a community is in equilibrium with its abiotic environment
Community all those species which share a habitat
(ecological) Succession progression of stages by which a group of organisms within a community reaches a state of equilibrium or climax
Ecosystem community of different species interacting with each other and the abiotic environment
Feedback response by a system to any change in its inputs
Food chain simple model showing energy flows through an ecosystem
Food web matrix of food chains
Habitat place or area in which a population lives
Hydrosere community adapting to an aquatic habitat
Net primary productivity (NPP) rate at which all the plants in a system produce useful energy, measured in kg/m^2/yr
Nutrient cycle circulation of minerals around an ecosystem
Nutrient flow transfer of nutrients between stores within the nutrient cycle
Plagioclimax succession an ecological succession determined or shaped by human intervention
Population all the members of a species living in an area
Primary succession plant communities occupying a site previously unvegetated
Secondary succession plant communities developing on sites that have formerly been vegetated
Sere succession from pioneer to climax community
Species group of organisms resembling one another in appearance, behaviour, chemical processes and genetic structure
System model showing how parts of the 'real' world function and interact; may be **open** or **closed**

Trophic level group of organisms having the same method of feeding or way of obtaining its energy
Trophic pyramid (also called **Energy flow pyramid**) diagrammatic way of displaying energy flows between trophic levels
Xerosere community that develops on dry land

1 Introduction to Systems

The natural world is very complex. Until the 1960s geography was concerned chiefly with describing 'how things were'. Then, in an attempt to understand the complex interrelationships among processes on the earth's surface, geographers introduced the use of **models**. Models are simplified representations of reality, in much the same way as a model train is a scaled-down version of the real thing. One type of model is a **system**. A system is a way of identifying an aspect of reality, known as a **unit** (e.g. a river valley), understanding the relationships between its separate components and then investigating how the unit interacts with the wider environment. A systems model is usually displayed as a **flow diagram** that, in its simplest form, can be shown as a **black box system** (see Figure 1.1).

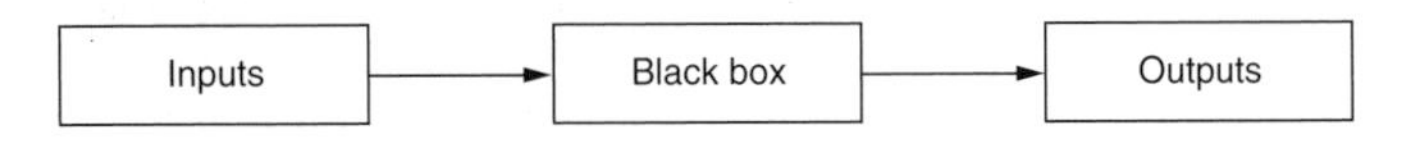

Figure 1.1 A black box system.

Such models allow us to identify **inputs** to a system (i.e. the entry of energy and/or matter) and **outputs** from the system (i.e. the mass, energy or change of state which leaves the system). However, the 'black box' (named because we cannot see into it) does not allow us to examine the **processes** which operate within the system. A more sophisticated model needs to incorporate information about what is happening internally and might look more like Figure 1.2 where:

- a **store** is a part of the system which can hold energy or matter
- a **transfer** is part of the system which redistributes energy or matter from one point to another
- a **flow** is any movement within the system.

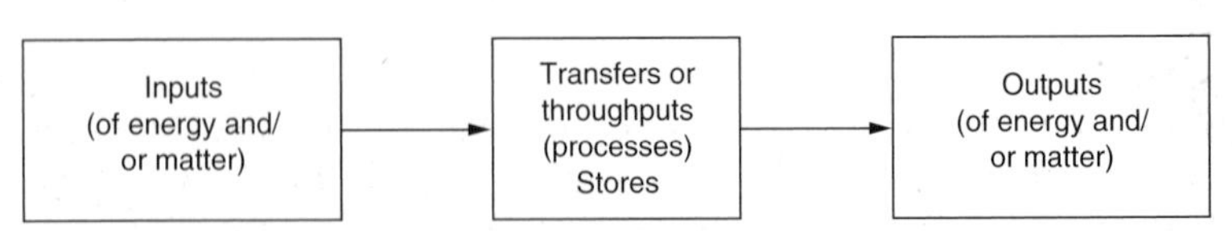

Figure 1.2 A typical systems diagram.

Systems can be classified as being:

- **Closed**: where there are inputs and outputs of energy but not of mass (or matter). Earth itself is often considered a closed system because it receives energy from the sun and loses heat into space but there is no transfer of matter between the universe and the planet. The global **hydrological cycle** is also a system of this type.
- **Open**: where there are inputs and outputs of both energy and matter. Such a system interacts with other, co-existing systems as well as with the surrounding environment.

When a system's inputs and outputs are balanced, it is said to be in a state of **dynamic equilibrium**. If one element in a system changes as a result of an outside influence it upsets the balance and affects other components in the system. This process is called **feedback** and may be either positive or negative.

Positive feedback occurs when the change has a 'snowball' effect with change becoming greater and greater. This moves the whole system away from equilibrium. **Negative feedback** occurs when the system acts to lessen the effects of the initial change and the processes within the system then work to restore the balance or equilibrium of the entire system (see Figure 1.3).

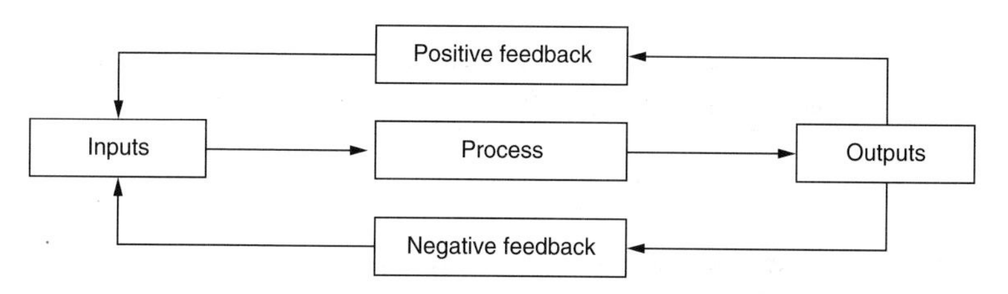

Figure 1.3 Positive and negative feedback loops.

2 Introduction to Ecosystems

A systems approach to studies of the living environment was adopted by biologists and environmentalists long before geographers incorporated it into their way of thinking. The unique contribution of **ecosystems** to our understanding of the natural world is due to its focus on the *interactions* between living organisms and their environment. In the past, the importance of these interactions was often overlooked and ecological disasters have occurred such as the 'Dust Bowl' of the 1930s in the USA.

Ecology (the biological study of the processes operating within ecosystems) has its own specialised vocabulary, and biogeographers have borrowed many of its key terms. Ecology is the study of **organisms**. An organism is any individual form of life; groups of similar organisms are known as a **species**. All of the members of a species living in an area are known as a **population**. The physical space, region or area in which a population lives is called a **habitat** and all the different species sharing a population's habitat are known as a **community**. An ecosystem is created when we consider a community of species interacting with each other and the non-living (**abiotic**) environment of energy and matter. All of earth's ecosystems make up the global component which biogeographers call the **biosphere** – although other scientists prefer the term **ecosphere**.

Not all ecosystems are natural; reservoirs, agricultural fields and garden ponds are all examples of ecosystems created by human intervention within the natural world. The size of an ecosystem is determined entirely by the requirements of the researchers studying it. For example, an ecosystem may comprise a single tree, a woodland or an entire forest stretching across thousands of kilometres, depending on the focus of the study being undertaken.

There are few clear-cut boundaries between different ecosystems. Even the apparently clear interface between land and sea is far from being a distinct boundary; this is partly because some species of animal have adapted to function equally well both in the sea and on land and also because of the formation of ecosystems such as salt marshes which form **transition zones (ecotones)** between land and sea. Ecotones are often richly populated habitats, supporting communities from both adjacent ecosystems as well as habitat-specific species.

Many ecosystems are extremely fragile; in order to protect them we now recognise the importance of having a detailed understanding of how they function and interact so that we can predict the likely outcomes of human initiatives and interferences, and act to reduce those which are damaging.

Climate patterns are the most important factor in determining which organisms can survive in a particular type of habitat. Certain animal species (most notably humans) have been able to **adapt** remarkably well to varying climatic conditions, whereas other species (particularly plants) have adapted slowly, over long periods of geological time, to live in very specific habitats and are now restricted to a particular climate and/or soil type. This tendency for plant species to be environmentally sensitive has led biologists to divide the earth's land surface (the biosphere) into large regions typified by distinctive climate and plant-life forms; such regions are called **biomes** (see Chapter 2). Within every biome there may be a range of ecosystems, reflecting plant **adaptations** to local soils, drainage, relief and micro-climates (i.e. local variations in temperature, moisture and light conditions).

The **hydrosphere** (earth's surface water) is sub-divided into **aquatic life zones** instead of biomes, the key factor being water salinity instead of climate. Lakes and streams make up the **freshwater life zone**, while estuaries, coasts, seas and deep oceans comprise the **marine life zone**. Both life zones support a range of ecosystems appropriate to the specific local environmental conditions.

In order for any terrestrial or marine ecosystem to be **sustainable** over time it must possess both the energy and the nutrients necessary to support its resident organisms, as well as the resources to dispose of and recycle their waste products. The living part of an ecosystem is known as its **biotic component** and the individual organisms residing there are often referred to as **biota**. The non-living parts of the ecosystem (e.g. solar energy, water, air and nutrients) are collectively called the **abiotic environment**. All ecosystems must comprise both biotic and abiotic components. The biotic organisms within ecosystems are categorised according to the way they obtain their food. **Autotrophs** (self-feeders) are green plants with the ability to produce sugars and other food compounds directly from abiotic nutrients via **photosynthesis**. During this process, chlorophyll converts solar energy, carbon dioxide and water into chemical energy such as glucose and oxygen. The oxygen by-product of photosynthesis is essential for maintaining most animal life on earth. In the hydrosphere, plants and algae are the main autotrophs in both freshwater and coastal environments. However, in deep oceans, **phytoplankton** are the dominant autotrophs. Autotrophs are often referred to as **primary producers**; all other organisms are known as **consumers** because they must consume other organisms in order to gain energy. Consumers (known more correctly as **heterotrophs**) may be sub-divided into:

- **Herbivores** (also known as **primary consumers**): these eat only primary producers, i.e. plants.
- **Carnivores** (meat eaters): these may be secondary or tertiary consumers. **Secondary consumers** eat herbivores while **tertiary consumers** eat other carnivores.
- **Omnivores**: these are meat and plant eaters; most omnivores are hunters.

There are also several other groups of consumers that, while fitting into the above categories, are recognised separately because they fulfil quite different roles within the recycling process:

- **Scavengers** feed on organisms killed by others or which have died from natural causes.
- **Detritivores** live off the waste products of other organisms.
- **Decomposers** (mainly consumers such as bacteria and fungi) complete the recycling of organic materials by breaking them down and releasing the resultant inorganic compounds back into the soil and water – where they become available once again as nutrients for the primary producers.

Energy flows through an ecosystem via **food chains** and **food webs**. In any ecosystem the origin of all its energy is sunlight. Sunlight is converted into energy by plants through the process of photosynthesis. As energy moves upwards through a food chain, much of it is lost – often as heat given out to the atmosphere. The remaining energy passes to the animals that eat the plants, and then to other animals that consume the plant-eaters. In its most simple form, a plant-to-animal food chain might look like the representation in Figure 1.4.

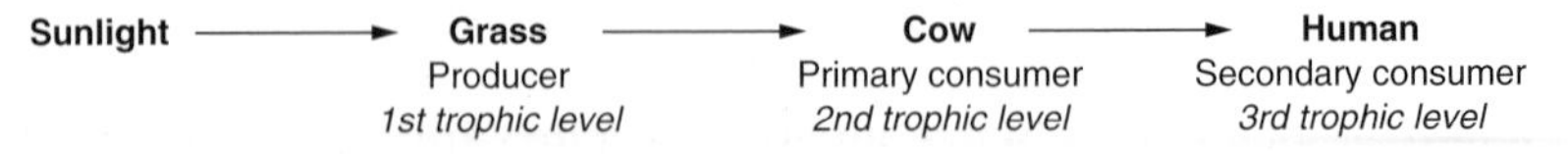

Figure 1.4 A trophic level food chain.

Biologists refer to each level within a food chain as a feeding level or **trophic level**. In the example given in Figure 1.4 there are three trophic levels.

In most ecosystems, the daily flows of energy are far more complex than simple food chains suggest; many animals are components in a large number of food chains and this creates networks of interconnected food chains that may be represented by a food web such as those in Figure 1.5.

The flow of energy within any ecosystem is a two-way process – up the food web and back down it – as energy is recycled for the 'next round' of the cycle. Recycling is usually shown as two interconnected food webs:

- **The grazing food web**, through which energy travels to the top carnivores (see Figure 1.6).
- **The detrital food web**, which represents the recycling of energy through organic waste materials (see Figure 1.6).

Much energy is lost as it flows through a food chain (or a food web), most of it to the environment as heat. Some residual energy passes straight through the animal unused and is excreted as waste. Often, as little as 10% of the total energy intake is used by the consumer; this energy is digested and converted into the organism's bodily material. The amount of energy actually available at each trophic level declines as the energy travels up the food web. Energy loss is often as great as 90% at each trophic level. This means that if plants capture 1000 units of energy, only 100 units are likely to be available to the herbivore(s) consuming the plant, and a mere 1% of the original energy reaches the carnivore at the next level in the food web. Such transfers and losses are often represented diagrammatically as a **trophic pyramid** or as an **energy flow pyramid** (see Figure 1.7b).

As a result of such a high rate of energy loss, it is very rare for food webs to reach the fifth trophic level; three or four levels are the more usual maximum. Beyond this level, there is so little residual energy left that the top carnivores find it almost impossible to support their own existence.

The total weight of all the dry organic material at any single trophic level in a pyramid is known as its **biomass**. Scientists calculate **dry weight** simply because the water which all organisms contain is of no use as either a nutrient source or an energy source. Biomass is

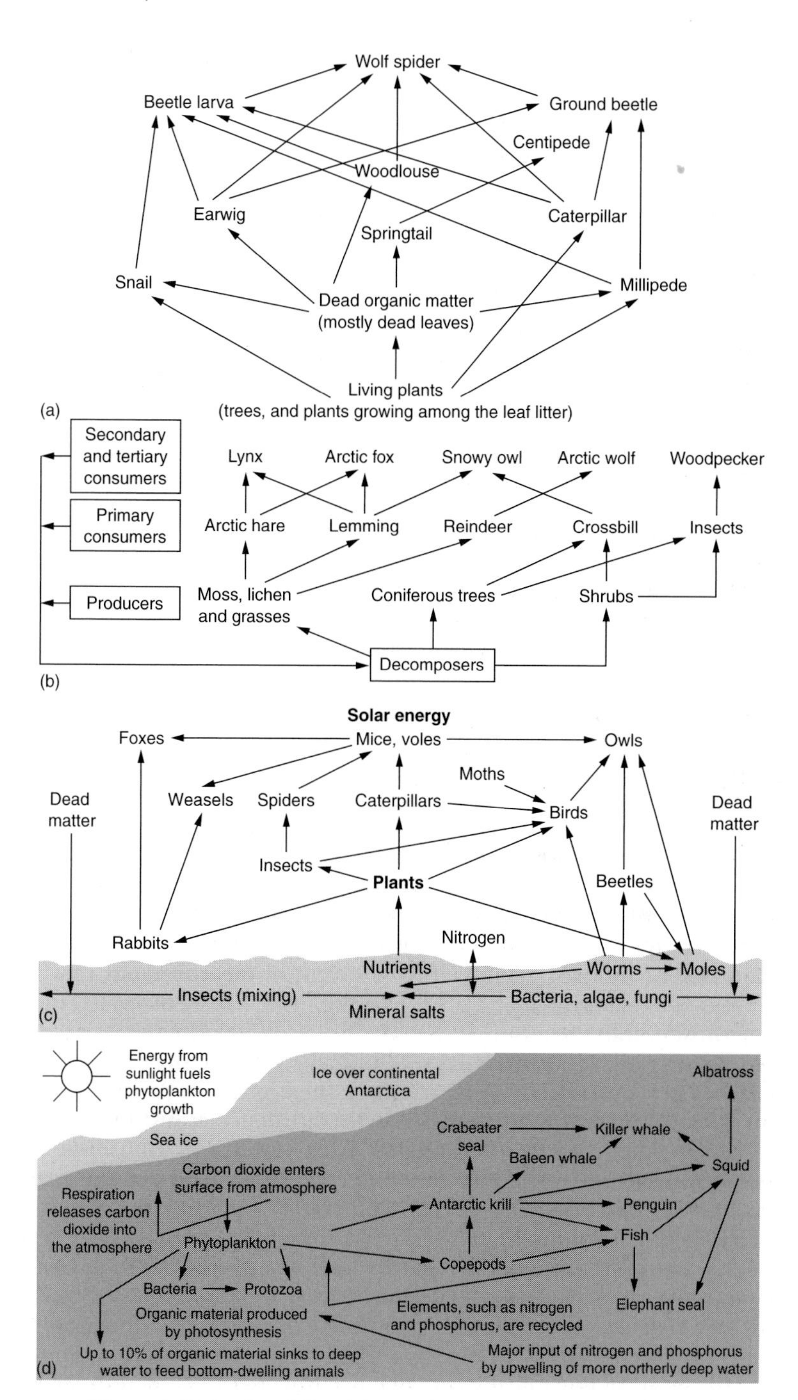

Figure 1.5 Food web for (a) minibeasts in soil and leaf litter, (b) a boreal forest system, (c) a deciduous forest ecosystem and (d) the Southern Ocean aquatic life zone.

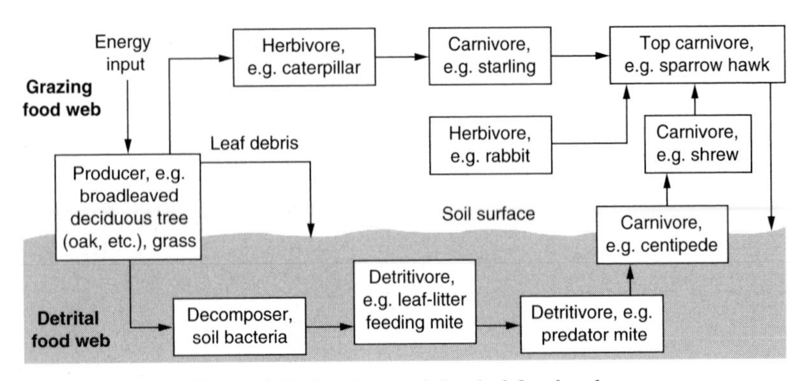

Figure 1.6 Grazing and detrital food webs.

used by scientists as a means of representing the chemical energy stored at each trophic level and this can be represented diagrammatically in pyramid form (although biomass pyramids are not always 'triangular' in shape). For example, the **biomass pyramid** for a typical ocean has a small base (representing its primary producers) but a large mass for primary consumers (see Figure 1.7a).

The rate at which any ecosystem converts solar energy into chemical energy is known as the system's **gross primary productivity** (GPP). However, such a measure is of little practical use as all of the energy produced by plants can never be made available to the primary consumers. This is because some of the energy that every plant produces must be used to sustain the producer. The residue available to consumers is the **net primary productivity** (NPP).

NPP is usually measured in units of energy (e.g. kilocalories) or mass (e.g. grams) per unit of area per year; such measurements will, therefore, have units such as $kcal/m^2/yr$ or $g/m^2/yr$. Where large quantities of biomass are involved, the unit of measurement may be expressed as kilograms or tonnes (i.e. $kg/m^2/yr$ or $t/m^2/yr$). There are considerable variations in NPP between different types of ecosystem or biome (see Figure 1.8). Tropical rainforests, estuaries and marshes are especially productive, but their primary productivity is not directly accessible to humans as food; estuarine and marsh vegetation is also inedible to humans. Most rainforest nutrients are stored within the vegetation rather than the soil, which is the main reason why deforested areas used as arable land become infertile within only two or three seasons. Not surprisingly, tundra regions, deserts and deep oceans have particularly low overall NPPs per unit area; the impressive total global productivity of oceanic areas is due to the fact that oceans cover 71% of the earth's surface rather than a high unit/area output. In direct contrast is the considerable annual contribution of the rainforests, which cover only 2% of the earth's surface.

In addition to energy flows, **flows of nutrients** take place within ecosystems. All organisms need nutrients. Nutrients such as nitrogen,

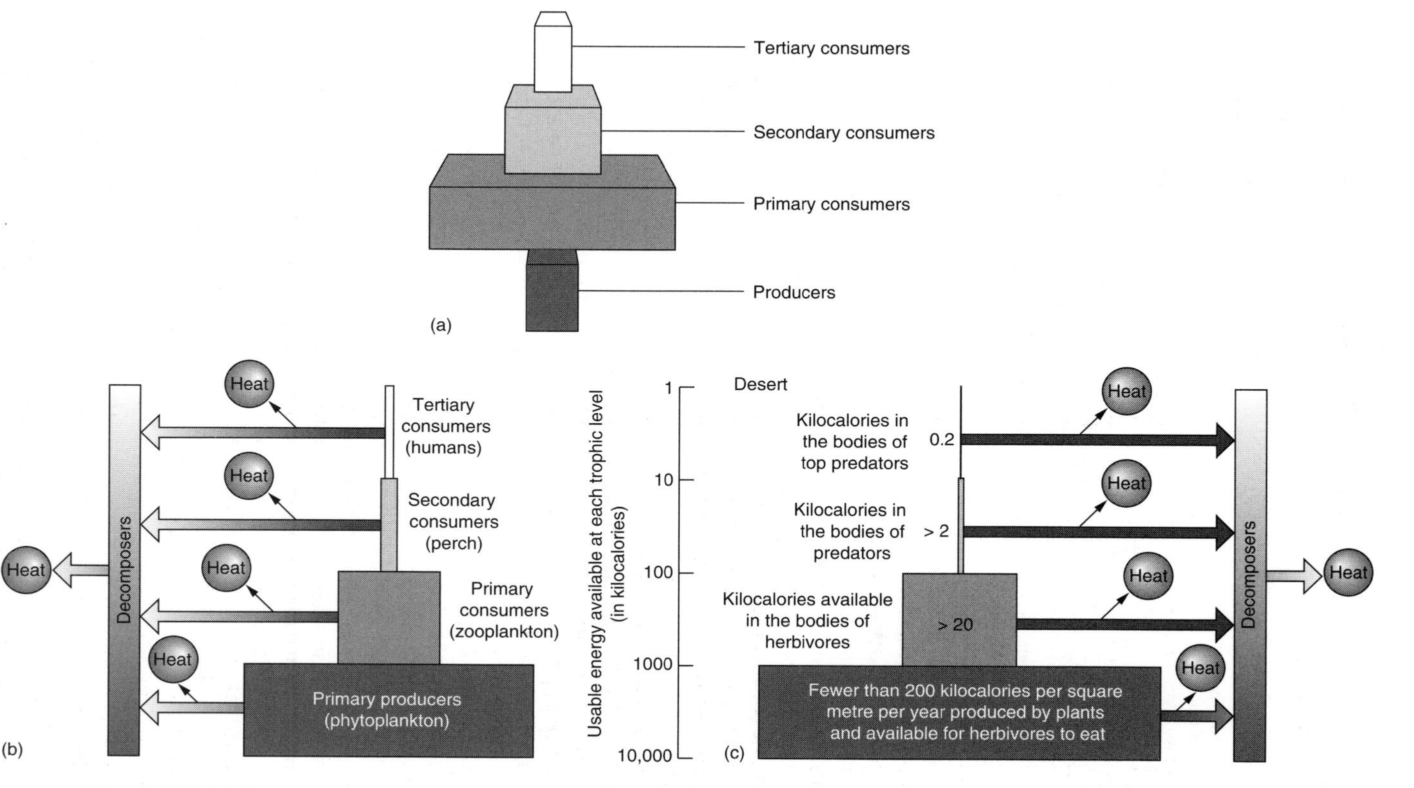

Figure 1.7 (a) A typical oceanic biomass pyramid. (b) A trophic pyramid. (c) A typical hot desert biomass pyramid.

Ecosystem type	Area (10^6 km^2)	Net primary productivity per unit area (g/m^2/yr)		World net primary production (10^9 t/yr)	World biomass (10^9 t)
		Range	Mean		
Tropical rainforest	17.0	1000–3500	2200	37.4	765
Temperate deciduous forest	7.0	600–2500	1200	8.4	210
Woodland and shrubland	8.5	250–1200	700	6.0	50
Boreal forest	12.0	400–2000	800	9.6	240
Savanna	15.0	200–2000	900	13.5	64
Temperate grassland	9.0	200–1500	600	5.4	14
Tundra and alpine	8.0	10–400	140	1.1	5
Desert and semi-desert scrub	18.0	10–250	90	1.6	13
Cultivated land	14.0	100–3500	650	9.1	14
Swamp and marsh	2.0	800–3500	2000	4.0	30
Open ocean	332.0	2–400	125	41.5	1.0
Algal beds and reefs	0.6	500–4000	2500	1.6	1.2
Estuaries	1.4	200–3500	1500	2.1	1.4

Figure 1.8 A global net primary productivity table.

carbon and phosphorus are circulated constantly throughout an ecosystem and can be absorbed by plants from the air (in gaseous form) or from the soil (as soluble salts). Animals ingest nutrients by consuming plants or other animals. The processes of death, decay and decomposition serve to release the nutrients stored in plant and animal tissue and create a new pool of nutrition ready to be recycled through the plants of the ecosystem. This process of nutrient recycling is usually modelled as shown in Figure 1.9(a) and is known as the **nutrient cycle** (or the Gersmehl Model). The model shows both 'stores' and 'transfers' within the recycling system. Nutrients may be stored within the litter layer, the soil or the biomass itself. The size of the 'circle' in such diagrams represents the percentage or proportion of the nutrient total stored in that component of the cycle. The arrows show both the direction of transfer and (by their width) the proportion of the total amount of nutrient transferred in this way. Figure 1.9(b–i) shows the nutrient cycles for the major biomes that are examined later in Chapters 4 and 5.

The transfer of nutrients along the 'uptake' pathway from the soil to the biomass represents uptakes of nitrogen, carbon and phosphorus (see Figure 1.10a–c) from the soil. The 'fallout' path represents the death of plants and animals and the addition of nutrients to the litter store. The 'decay' path from the litter store to the soil represents the decomposition of litter into humus and the return of

Figure 1.9 (a) A theoretical (Gersmehl) model of nutrient cycling. This model represents the storage and flows of nutrients in the ecosystem, not the size of the biomass and litter. In (b)–(i) the circle size is proportional to the total amount of nutrients in the system and arrow widths indicate the quantity of **nutrient flow** expressed as a proportion of that stored in the sources pool. B, biomass; L, litter; S, soil. After Tivy (1982).

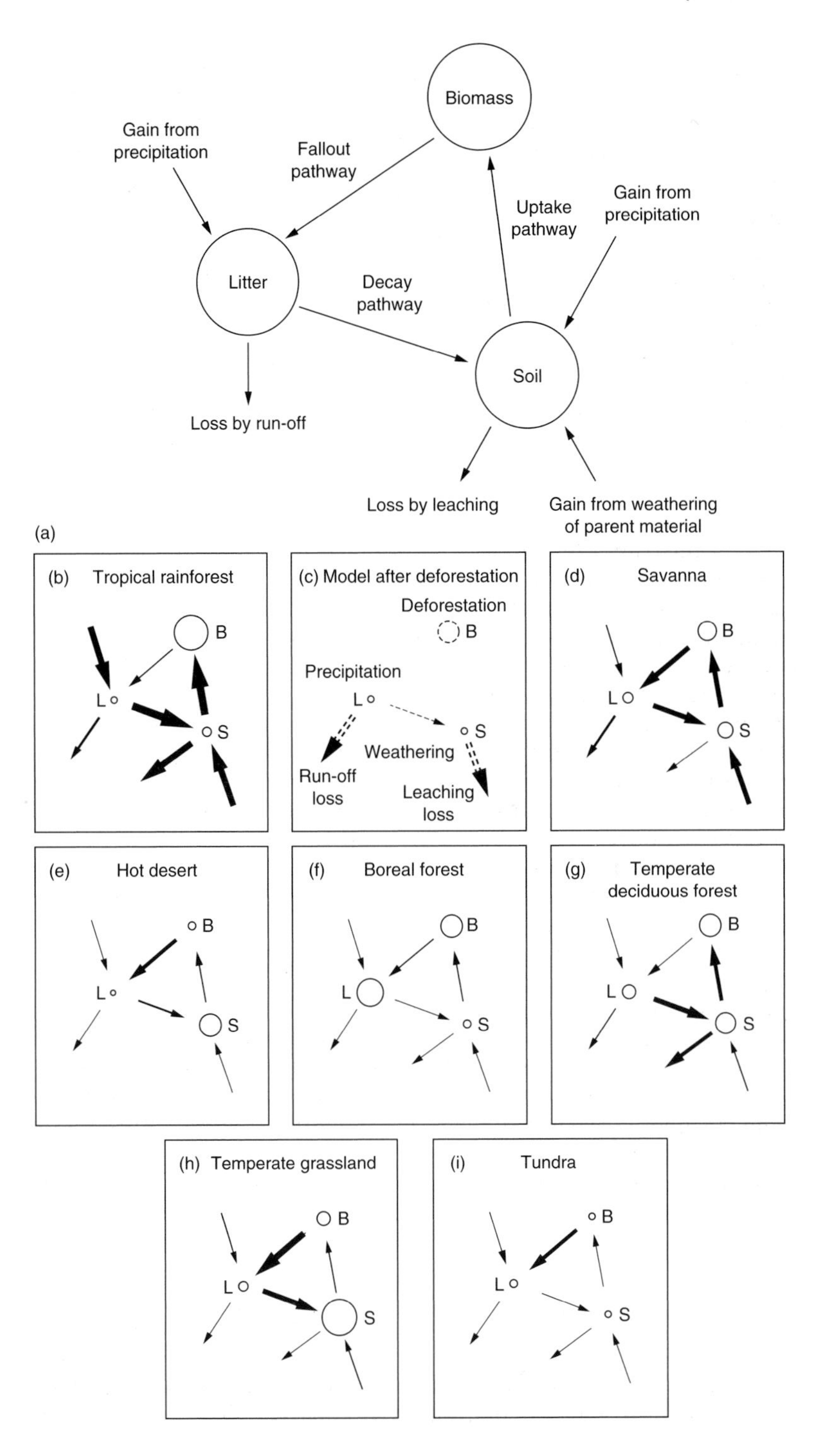
Biomass
Gain from precipitation
Fallout pathway
Uptake pathway
Gain from precipitation
Litter
Decay pathway
Soil
Loss by run-off
Loss by leaching
Gain from weathering of parent material
(a)
(b) Tropical rainforest
B
L
S
(c) Model after deforestation
Deforestation
B
Precipitation
L
S
Weathering
Run-off loss
Leaching loss
(d) Savanna
B
L
S
(e) Hot desert
B
L
S
(f) Boreal forest
B
L
S
(g) Temperate deciduous forest
B
L
S
(h) Temperate grassland
B
L
S
(i) Tundra
B
L
S

nutrients to the soil. The recycling of nutrients is an open system as nutrients may be added to the cycle and/or lost from it. Additions may be natural (e.g. animals may move into an ecosystem, leaves may blow into an area or chemicals may be added to soil by weathering of the parent material) or human (by adding organic or inorganic fertilisers, by planting crops or trees); similarly losses from the system may be either natural or be brought about by human interventions. Natural losses include nutrient loss via leaching (see page 44) and animal movements out of the ecosystem, while human interventions may include deforestation, overcultivation/grazing, harvesting of natural vegetation for fuel and crop harvesting.

All life-forms require constant supplies of nitrogen, carbon and phosphorus. The **carbon cycle** and the **nitrogen cycle** are described as being **atmospheric cycles** because both elements (carbon as carbon dioxide) are available within the atmosphere. The recycling of phosphorus is important because it is a critical part of DNA; it is also necessary for respiration and is present in bones, teeth and shells. Unlike nitrogen and carbon, phosphorus does not circulate in the atmosphere because it cannot exist in a gaseous form; the **phosphorus cycle** is, therefore, a **sedimentary cycle** (see Figure 1.10b). All sedimentary cycles are slow and recycling of phosphorus in soil can be especially so; this can result in phosphorus deficiencies which inhibit plant growth if they continue over long periods of time.

One of the most notable features of any community or ecosystem is that it exists in a state not of stability, but of constant change. Such change is gradual and is the inevitable result of the ecosystem perpetually seeking balance. Environmental conditions are constantly subject to modification and ecosystems reflect these changes over time. Individual organisms and species have to adapt to such change or risk extinction. Therefore, a stable ecosystem is not a static one; it continually adapts and rebalances itself in ways which maintain it in a constant state of flux. Plant adaptation is often the first discernable indicator of such change – unless the ecosystem is responding to some sudden, catastrophic event. Adaptation occurs when any organism develops beneficial mutations that allow it to cope more effectively in changing environmental conditions and to produce offspring with the same adaptive traits. In plants, this often enhances the individual's ability to compete for light, space and nutrients. Vegetation adaptations usually have a knock-on effect to the animal communities sharing the habitat. The gradual adjustment process is called **ecological succession** (or **community development**).

a) Succession

i) Primary succession

Ecological succession is of two kinds. **Primary succession** involves the development of biotic communities in areas lacking soil (or lacking

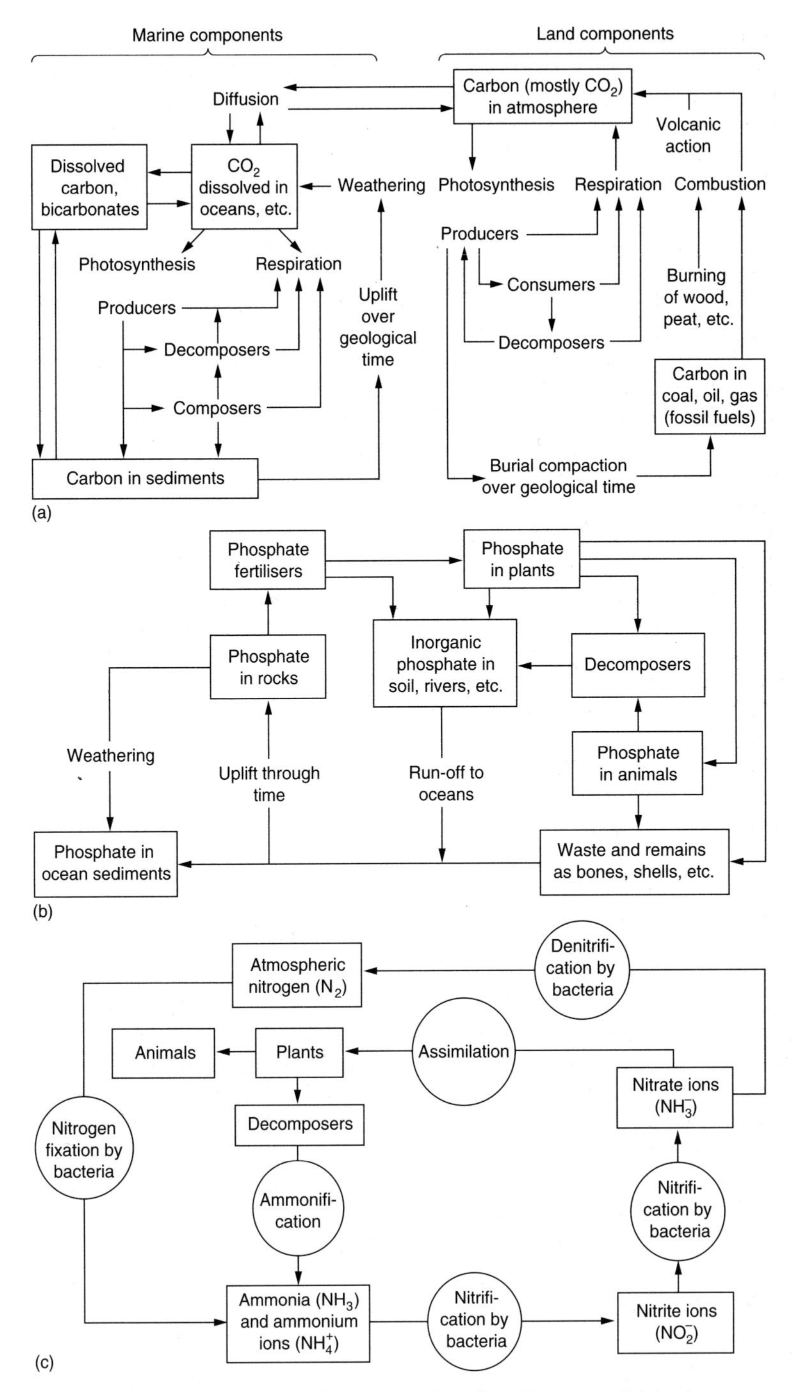

Figure 1.10 Global cycles: (a) carbon, (b) phosphorus and (c) nitrogen.

bottom sediment in water). Such environments may include recently cooled lava flows, freshly cut quarry faces and new garden ponds. On land, plants cannot survive without soil; therefore, **pioneer species** which colonise newly exposed surfaces are always soil-forming species such as mosses and lichens. Such species are able to take the nutrients that they need directly from the rock surface. As soil begins to form, the area will be colonised by species such as bacteria, fungi and insects which, when they die, will add to the organic content of the embryonic soil. Eventually, short grasses, herbs and ferns are also able to move in. Such colonising species are known collectively as **early successional species**. It may take several hundred years for the soil to become deep and fertile enough to support **mid-successional species** such as taller grasses and shrubs. Eventually, however, trees will begin to colonise the area and grow towards maturity. As the range of plant species increases within an area, so too do the animal species co-existing alongside them; animals can only move in after their sources of food have become secure, and early successional animal species are always small herbivores. Large **grazers** often arrive much later, long after small predator carnivores have become well established. Provided that primary succession is not interrupted by natural or human interventions, **late-successional species** ultimately colonise the habitat and reach maturity. The originally barren area will eventually support a stable, complex and mature community.

Nowadays, it is increasingly rare to find undisturbed primary succession communities. Even in our most remote 'wilderness' areas, past interventions by human activity have disturbed the ecosystem in some way.

ii) Secondary succession

Secondary succession is far more common. This occurs where the original natural vegetation has been disturbed or wholly destroyed, but only where the soil has remained *in situ*. It is unusual for secondary succession to replicate fully the primary community that it replaces. This may be because there will have been subtle changes in the environment which means that slightly different species will now colonise the area. More likely, such changes will occur simply because the whole process is not starting again from scratch. As soil is already in place, the initial colonisation process does not have to be repeated; this will therefore create a different evolutionary process, which means an identical succession cannot redevelop in this habitat. If, on the other hand, the soil cover is damaged (e.g. by erosion), it is even less likely that the primary community will be able to re-establish itself. Changes to the soil's composition, depth and fertility will facilitate the development of a new and different succession, with a major knock-on effect to the other animal species moving into the habitat as vegetation cover develops.

iii) Climax communities

Traditionally, ecological succession has been viewed as a natural and orderly growth towards a **climax community**. Climax communities used to be regarded as the ultimate goal of the developmental process. They were typified by a stable, predictable community based on well-established plant species (often trees) that could sustain and replace themselves well into the future. However, recent research suggests that both primary and secondary successions are highly unpredictable processes. Studies of secondary succession show that earlier plant species rarely recolonise a habitat following clearance. Many of today's leading ecologists believe that even single, random or chance events can create major changes to ecological succession within habitats. They reject the idea of an 'ecological plan' under which stable, climax communities are produced; instead, they think that succession is the product of unique struggles between competing species for light, air, water and essential nutrients. Therefore, these scientists prefer the term **relatively stable community** (or **mature community**) rather than the traditional idea of climax community. Even so, this terminology still implies that the succession has been uninterrupted by human interference.

Today, such communities are very rare indeed. Primary or secondary successions halted or modified by human interference (intentional or not) or by natural events are known as **plagioclimax successions**. While a range of natural hazards can interrupt successional development, perhaps the most common arresting factor is fire. Fire may occur spontaneously (usually as a result of a lightning strike during a drought when dormant vegetation is dry) or may be anthropologically induced either with or without purpose.

Ecosystems that are affected naturally and regularly by fire often adapt to the event. Some vegetation (e.g. palm trees) becomes **pyrophytic**, while grasses and flowering plants adapt by growing from buds underground or from rhizomes, which protect the plant against permanent fire damage. Other vegetation (e.g. the jack pine) has adapted so that the reproductive cycle becomes dependent on a fire occurring; the cones are covered in a resinous material that needs the heat from a fire to melt it in order to release the seeds within. After a forest fire, soils are fertilised with deposits of ash. This ensures that the new seedlings begin life with a plentiful supply of nutrients. Where vegetation is removed by fire, the opportunity occurs for secondary succession to regenerate the area. This promotes **biodiversity** by increasing the number and range of habitats available to the new colonisers and may be the reason why tropical rainforest areas affected by slash-and-burn farming often display greater species diversity than the surrounding forest ecosystem. However, the suppression of fire appears to inhibit the seeding of pyrophytic species and tends to lead to increasingly catastrophic events if, or when, fire takes hold in such areas. Many ecologists now believe that where fires are a natural part of the ecological process, they should be allowed to occur spontaneously. Conversely, 'managed' fire used as a tool to advance human activity should be discouraged, because it introduces a control mechanism that is not part of the natural system.

b) Prisere development

The sequence of events across time that produces a mature community from a barren expanse of land or sea is known as a **prisere** and each individual phase of development towards maturity is referred to as a **sere**. Communities that develop on dry land are classified as **xeroseres**, while those adapting to water are known as **hydroseres**. Specialised xerosere communities developing in extremely dry conditions (e.g. on sand dunes) are sub-classified as **psammoseres** while those communities developing on land possessing water-retaining properties are sub-classified as **lithoseres**. In aquatic environments, the established sub-classification can be a little confusing because all water adaptations are known as hydroseres, but this term is also the sub-classification for all freshwater ecosystems. Communities adapting to saline conditions are sub-classified as **haloseres**. Both haloseres and psammoseres are, in fact, transitional seres in the development of mature land-based communities. For example, salt marsh communities inevitably dry out as the sea retreats, eventually becoming dry land. Salt traces are eventually washed from the soil and land-based plants are then able to colonise the 'reclaimed' land – ultimately generating fully mature xerosere communities. Figure 1.11 traces all of the above developments within dry and wet environments.

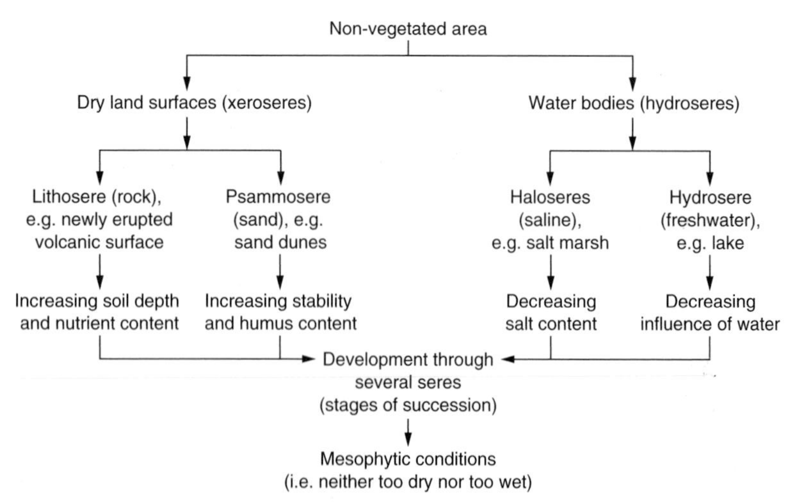

Figure 1.11 Prisere development.

Summary

- A system is a way of modelling reality that allows the separate components and, importantly, the interactions between individual components to be examined. A system may be described as being either 'open' or 'closed'.
- The use of systems models allows identification and examination of inputs, outputs, stores and flows (transfers), together with the processes themselves.
- When inputs and outputs are balanced, a system is described as being in equilibrium; any change, however slight, upsets this balance and creates feedback. Feedback may be positive (if it enhances the introduced change) or negative (if it reduces the change and works to rebalance the system).
- Geographers use the concept of an 'ecosystem' to study communities of plants and animals sharing a habitat, and to gain understanding of the interrelationships between them and their abiotic environment. Boundaries between ecosystems are rarely fixed; the transition zones between ecosystems are known as ecotones.
- Organisms adapt to specific conditions within their habitat. Adaptation takes time and a variety of forms: humans are the most adaptable organisms while plants are the least adaptable to change.
- Organisms are categorised according to how they are supplied with energy. Autotrophs produce food compounds from the abiotic environment, but all other organisms are consumers.
- Energy flows around an ecosystem through a network of food chains and webs. As energy moves upwards through these networks, much of it is lost – often as heat.
- Flows of nutrients also circulate within ecosystems. These may be represented by a Gersmehl Model, which shows stores and transfers of nutrients through the biomass, leaf litter and soil.
- Primary succession involves the development of biotic communities in areas that lack soil cover. The sequence of events leading to the establishment of soil and stable, mature (climax) communities is known as a prisere. Each stage of succession is referred to as a sere.
- Secondary succession occurs when vegetation has been disturbed or destroyed but soil cover remains. Such succession is unlikely to replicate the original mature vegetation cover and results in a plagioclimax community.

Student Activities

1. Explain the following terms: **a)** an open system, **b)** a positive feedback loop, **c)** a food web, **d)** a trophic pyramid, **e)** primary succession, **f)** adaptation, **g)** a plagioclimax community.

2. **a)** Outline the meanings of 'primary succession' and 'climatic climax (mature) vegetation', and then state the link between them.
   b) What is 'secondary succession'?
   c) Explain how human intervention may lead to the development of both secondary succession and plagioclimax succession.
3. Describe the ways in which a succession towards mature vegetation can be arrested by: **a)** natural events, **b)** biotic intervention and **c)** anthropogenic intervention.

2 Biomes

KEY WORDS

Actual evapotranspiration amount of evapotranspiration that actually occurs
Anthropogenic originating in human activity
Aquatic life zone water-based ecosystem
Biome one of the major sub-divisions of the earth's biosphere
Ecotone transition zone between adjacent ecosystems
Edaphic relating to the soil
Effective moisture net soil moisture available for use by vegetation
Effective soil moisture balance between precipitation inputs and the demand for water for evapotranspiration
Evapotranspiration the combined effect of evaporation and transpiration in moving water from the soil store back into the atmosphere
Formation one of the main sub-divisions of a biome
Growing season part of the year when mean temperatures are above 6°C
Potential evapotranspiration amount of evapotranspiration taking place if there was an unlimited water supply
Soil erosion removal of topsoil by wind, water, ice or human activity

1 Introduction

Biomes are the largest 'ecosystem' unit. They are identified by a shared, characteristic plant community adapted to the specific environmental conditions of the region. There is a strong positive correlation between global biome distribution and world climatic zones (see Figures 2.1 and 2.2) because climate is the most influential factor in determining natural vegetation distribution. Water-based units are more accurately known as **aquatic life zones**. Within such zones, salinity replaces climate as the crucial key to adaptation, although temperature, light and nutrients also play important roles.

Maps such as Figure 2.2 suggest that biomes are fixed regions with inflexible boundaries marking significant changes between adjacent plant communities. They also give the impression that totally uniform vegetation communities cover huge expanses of the planet. Nothing could be further from reality. Biomes are merely convenient classifications of dominant vegetation type, each characterised by variation as well as by similarities. Any sizeable forest, for example, may consist of a range of different trees, shrubs and flowering plants adapted to a range of localised environmental conditions. Biomes that span

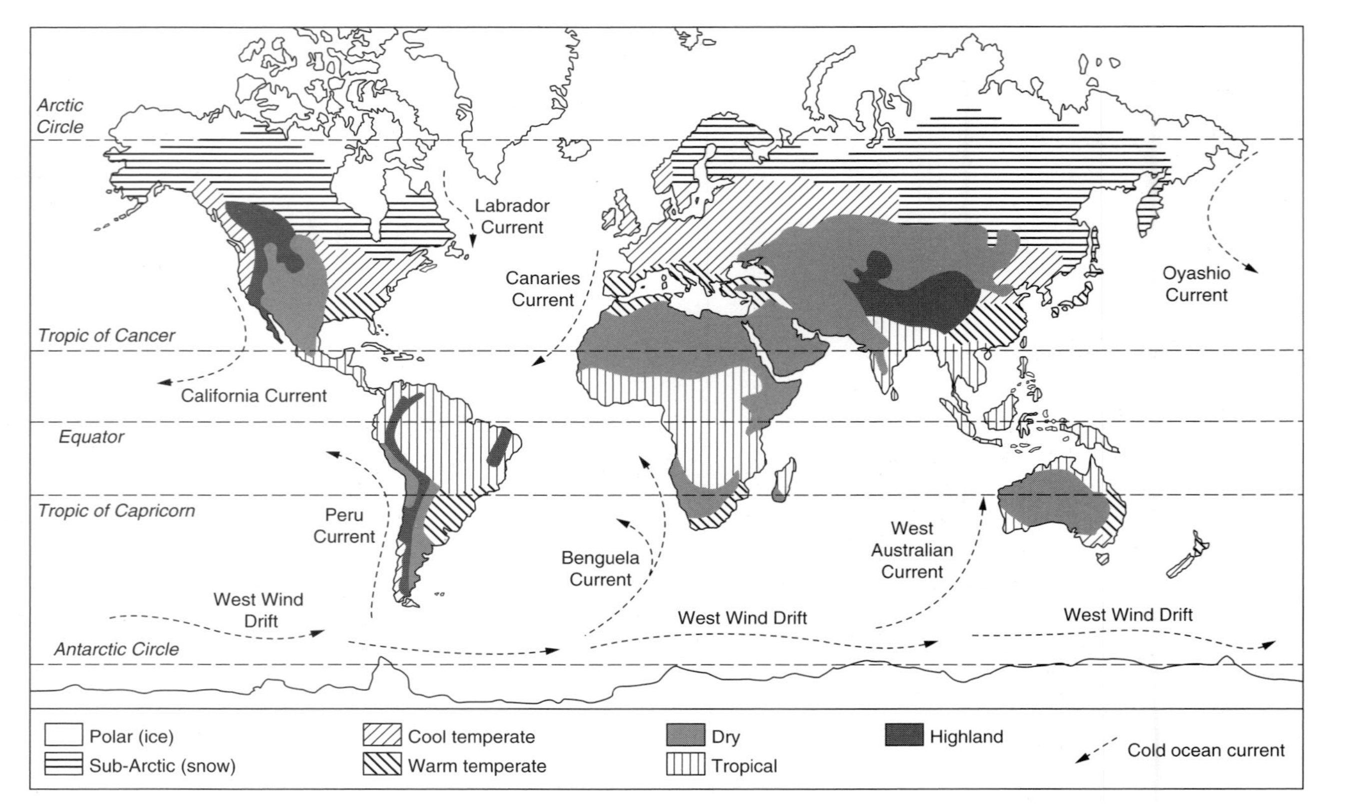

Figure 2.1 Global climate patterns and cold ocean currents.

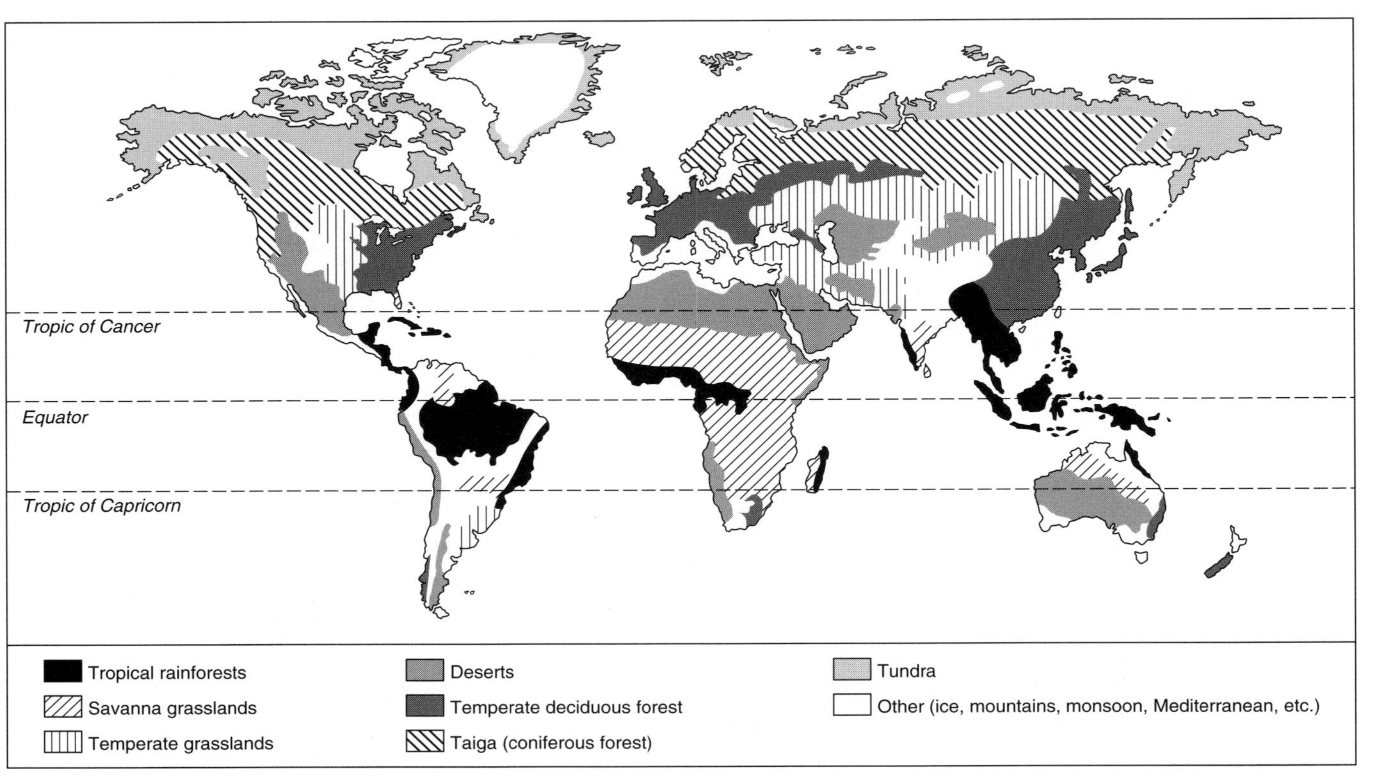

Figure 2.2 Global biomes.

continental areas may exhibit wide variations in vegetation across the region and may host many varied animal species located within specific habitats. One reason for this is that both within and between continents, soil types may vary considerably according to parent rock, glacial histories and local microclimates. Within any one biome, a wide variety of ecosystems is likely to develop in response to local **edaphic conditions** and also the degree to which human intervention has taken place. Surprisingly rich and dense **ecotones** often mark the transition between biomes and between neighbouring ecosystems. As a result, the earth's surface is not covered by uniform 'blankets' of identical communities but by 'patchwork quilts' of ecosystems/ecotones, with each combination exhibiting its own distinctive, locality based variations.

At the global scale, four natural factors interrelate to produce the biome distribution. These are **climate**, **topography**, soils and biotic factors, to which the role of human intervention in ecosystem development needs to be added.

2 Climate

Precipitation is the single most important determinant of vegetation type, be it forest, grassland or desert; coupled with temperatures and soil type, it influences whether a region's vegetation is tropical, temperate or polar.

a) Precipitation

The importance of precipitation extends far beyond annual totals; of greater importance is the distribution of precipitation throughout the year. For example, large quantities of rainfall in one season, followed by prolonged drought will favour one type of mature (climax) community, whereas precisely the same total of annual rainfall distributed evenly throughout the year will lead inevitably to the dominance of a totally different plant community. Even so, precipitation needs to be considered in relation to ambient temperatures because the demands of **evapotranspiration** are an essential consideration for plant successions. Where (and when) temperatures are high, evapotranspiration demands are also high and this is critical in determining whether or not the rainfall totals and distribution patterns are able to sustain vegetation growth. For example, in areas where rain falls all year round and temperatures are high, forest growth is possible; in equatorial areas, heavy rains throughout the year offset the high evapotranspiration demands of the high temperatures. Yet low precipitation totals in cold, high latitudes also support forests because the evapotranspiration demands on available moisture are significantly less in such locations. Where prolonged droughts (particularly summer droughts) coincide with high rates of evapotranspiration – as in the savanna grassland biomes (see Chapter 4), plants must be **xerophytic** (drought resistant) in order to survive. The relationship between precipitation and biomes is not simply about quantity and distribution but is about effectiveness.

Effective moisture is the term used to describe the net soil moisture available for the vegetation to use as and when it is required. **Effective soil moisture** reflects the balance between the input of precipitation (P) and the demand for water for evapotranspiration (eT). When precipitation exceeds evapotranspiration (i.e. when $P > eT$) then surplus moisture is stored in the soil and the **soil moisture budget** is described as being positive. When evapotranspiration exceeds precipitation ($eT > P$), plants will initially use the stored soil moisture, but once this has been exhausted, the soil moisture budget is described as being negative (see Figure 2.3). Plants colonising areas experiencing **negative soil moisture budgets** need adaptations in order to survive. Evapotranspiration rates are at their highest during the 'high-sun' (or summer) season, although this does not necessarily mean that evapotranspiration is low during the 'low-sun' (or winter) season, only that it is lower than at other times. When precipitation occurs, soil moisture may be recharged during times of lower water demand and in many (but not all) regions of the world, precipitation is sufficiently abundant across the year to replenish the soil moisture store at some point.

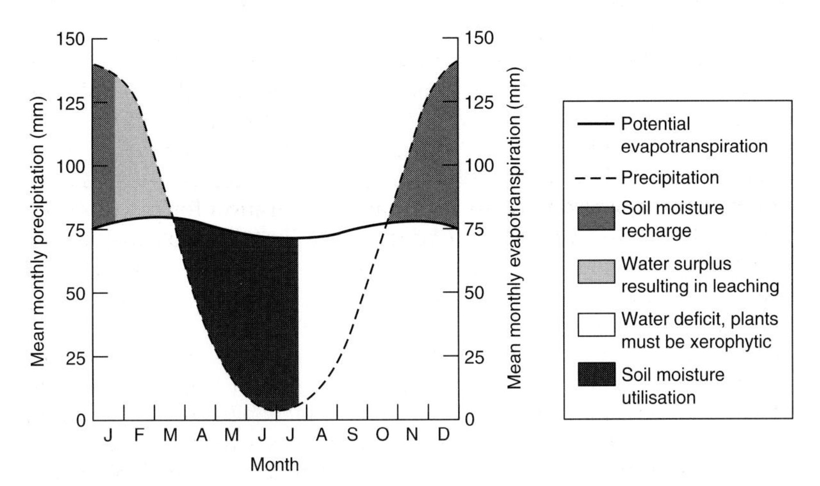

Figure 2.3 Soil moisture balance for African savanna south of the equator.

b) Temperature

In addition to influencing evapotranspiration, temperature affects plant adaptation and survival. All plants have maximum and minimum temperature tolerances within which growth can take place. This tolerance range varies quite widely between plant species but, generally, plants cease to function (i.e. produce chlorophyll) and so become dormant when the air temperature falls below 6°C; ideally, temperatures should exceed 10°C for effective photosynthesis to take place. The **growing season** is determined by the number of months (or weeks) when temperatures are high enough for plant growth to occur, i.e. are above 6°C. Where temperatures exceed 15°C throughout the year, there is the potential for a continuous growing season, although plants often exhibit distinct signs of stress when temperatures exceed 35°C. For most plants, the optimum mean annual temperature for growth is 25°C; beyond this temperature, the plants' water requirements increase greatly. Wherever winter temperatures fall below 6°C for up to 5 months of the year, trees adapt by shedding their leaves to protect themselves from frost damage. Where temperatures fall below 6°C for more than 6 months of the year, tree adaptation is to retain their leaves in order to maximise photosynthesis as soon as temperatures rise above the critical temperature. The influence of temperature on vegetation is not usually experienced in isolation, but operates in conjunction with other factors such as humidity, precipitation and light intensity. For example, well-established olive groves have the ability to survive periods of extreme cold provided that the weather remains dry; yet fairly short spells of cold, wet conditions can be extremely damaging.

c) Light

The availability of light determines the rate at which photosynthesis occurs. Light availability and its intensity vary considerably between places and at different times. Both factors are influenced by latitude, season, local relief, climate (especially cloud cover) and the proximity of nearby plants. As light decreases, fewer plants are able to exist and so they become more widely dispersed. Light quality is also an important factor and high ultraviolet light levels seem to be a significant reason why the range of plant species is greatly reduced in mountainous areas.

d) Wind

Atmospheric movements also affect ecosystem and biome development. Wind direction and strength influence both evapotranspiration rates and air temperatures. In colder climates, the wind-chill factor can lower the effective temperature by many degrees. Wind can also hasten the drying-out of soil in exposed locations, leading to a reduction in effective soil moisture and a marked increase in the potential for **soil erosion**. Permanently windy habitats invariably support grassland rather than tree cover, partly because grasses have adaptations allowing them to bend easily without damage to their internal structures.

3 Topography

Topography (or relief) can influence vegetation successions in several ways. Increases in altitude are closely correlated to decreases in temperature and so increased altitude tends to result in stunted plant growth, fewer plant species and a consequent reduction in protective soil cover. Relief may also increase an area's exposure to (or indeed, protect it from) heavy rainfall and strong winds. Gradient (or angle of slope) also affects soil depth, texture, acidity and drainage, with most steep slopes having thinner, but less waterlogged soils (see page 32).

Aspect can be important locally. Within the Northern Hemisphere, south-facing slopes are more favourable locations for plant growth as the orientation increases access to sunlight, leads to higher temperatures and reduces susceptibility to frosts.

4 Edaphic Factors

The equilibrium between soils and vegetation is extremely delicate. The availability of organic and inorganic matter within the soil is important for the continued well-being of the entire biome. While biomes may be characterised by specific soil types, considerable variations can occur as a result of local differences in soil and/or underlying parent rock. Britain provides an excellent example of such variation; although the regional classification is 'temperate deciduous forest', the mature vegetation across large tracts of the British countryside is predominantly grassland and heathland owing to underlying chalk and limestone giving rise to local variations in soil type. Vegetation cover is also affected by local variations in the soil cover such as texture, structure, acidity and depth as well as by water retention, oxygen and nutrient content.

5 Biotic Factors

The most important **biotic factor** affecting biomes is the interplant competition for light, root space and water. Such competition is often fierce in the lower latitudes where vegetation cover is particularly dense and a greater numbers of species is involved. However, **grazers** and **browsers** also exert a considerable influence on vegetation development. Herbivores and primary consumers are frequently responsible for pollination and seed dispersal, as well as the close-cropping of the vegetation itself. Second and higher-order predators are influential by controlling the populations of herbivores and primary predators such as insects. Animal diseases are an additional factor as was clearly demonstrated by the unintentional introduction of myxomatosis to Britain in the 1950s. Myxomatosis is a viral infection of rabbits endemic only in South America. However, it was deliberately introduced into Australia in the 1950s as a means of dramatically reducing the rabbit population and was inadvertently transmitted to Britain at about the same time, decimating the rabbit population there as well. The unexpected rabbit cull led to record rates of grass growth, increased crop yields and successful sapling establishment.

6 Anthropogenic Influences

In addition to the above natural influences, the role of human activity should not be underestimated. Very few areas of the world support 'undisturbed' mature (climax) communities or biomes and in many places, particularly the major urban conurbations, the natural biome has been either totally destroyed or very comprehensively manipulated in order to meet the perceived needs of the local human population. Human interventions in the natural world may be categorised as being of three distinct types:

- In the atmosphere, the delicate balances of carbon dioxide and ozone within the lower and upper atmospheres respectively are being destroyed, leading to temperature rises and increased ultraviolet radiation. The current rates of sulphur dioxide and nitrogen oxide emissions are devastating large areas of forest biomes through the effects of **acid rain** (see Chapter 7).
- On land, our pursuit of economic activities has led to rapidly increasing urban sprawl as well as widespread stripping of natural vegetation to provide additional land for both building and large-scale agriculture. Subsidence, soil exhaustion, topsoil erosion, deforestation, fire and flooding are just some of the many factors currently responsible for biome modification, degradation and destruction. Even **reafforestation** programmes can bring about major ecological change, particularly where huge stands (clusters) of single species evergreens are planted to replace deciduous or mixed woodlands. Such 'new' species cannot host the previous animal communities as they do not replace the food supply and living space of the natural woodland. Furthermore, over a period of time, the natural soil cover of the area becomes affected by the input of acidic leaf litter and increased leaching and consequently degenerates, becoming unable to support the original forest succession even if recolonisation is attempted.

- Disturbance of the ecological balance has taken place owing to the widespread use of increasingly potent fertilisers, pesticides, herbicides and fungicides. Such disturbance, combined with **overgrazing** by animal herds, has led to changes in the inputs and outputs of the natural systems and created new, unnatural, feedback loops. Often, this feedback is positive, resulting in increasingly rapid movement away from equilibrium and accelerated rates of change.

Summary

- Climate patterns are the most important factor in determining adaptation. Large areas typified by distinctive climates and plant forms are known as biomes.
- Precipitation is the single most important determinant of the three vegetation types: forest, grassland or desert. Coupled with temperature and soil type, precipitation also determines whether vegetation is tropical, temperate or polar. Hence, there are nine primary biome classifications – see Figure 2.2.
- Temperature influences the rate of evapotranspiration, which in turn determines effective soil moisture – the moisture available for plant growth. Plants colonising areas experiencing negative soil moisture budgets must adapt in order to survive.
- Temperature determines the length of the growing season. Plants cease to function when temperatures fall below 6°C; ideally, temperatures should remain above 10°C for effective photosynthesis. If temperatures exceed 15°C throughout the year, the growing season is continuous. Where temperatures are below 6°C for up to 5 months, trees adopt a deciduous habit; where more than 6 consecutive months are below 6°C, trees become 'evergreen' in order to maximise photosynthesis opportunities.
- Relief can affect plant growth because although temperatures decrease with increasing altitude, precipitation is greater. Strong winds, edaphic conditions and waterlogging are also influenced by relief. Aspect may be an additional, locally important factor.
- Plant adaptation is also affected by a range of edaphic factors, which are discussed at length in Chapter 3.
- Animals also affect vegetation cover and adaptation, many playing a significant role in pollination and seed dispersal, while grazers and browsers crop vegetation cover. This is an important factor in inhibiting tree colonisation/regeneration.
- Human intervention affects nearly every ecosystem on earth. Such interference may take the form of atmospheric pollutants (e.g. acid rain), biome modification (e.g. through the removal of vegetation cover) and ecological disturbance (e.g. usage of pesticides), all of which introduce positive feedback loops, accelerating change away from natural equilibrium.

Student Activities

1. Explain the following terms: **a)** ecotone, **b)** edaphic, **c)** effective soil moisture, **d)** evapotranspiration, **e)** growing season, **f)** negative soil moisture, **g)** soil erosion, **h)** soil moisture recharge.
2. 'Biomes reflect the pattern of world climate and soils'. To what extent do you agree with this statement?

3 Soils

KEY WORDS

Calcification laying down of calcium-rich material in the soil profile
Catena sequence of soil types along a slope that are distinct from each other because of variations such as drainage, leaching and mass movement
Cation positively charged ion
Cation exchange process by which cations are exchanged between plant roots and clay/humus particles
Cheluviation form of eluviation occurring within the 'O' horizon by which iron and aluminium are translocated by organic acids through the soil profile
Eluviation washing-down of nutrients through soil horizons
Ferrallitisation enrichment of soil by iron and aluminium oxides
(soil) Horizon distinctive stratum or layer within a soil profile
Humus fully decayed organic matter
Illuviation laying down of material in the lower horizons of the soil profile
Leaching removal down through the soil profile of organic materials and minerals in solution
Loam soil characterised by having equal proportions of sand, silt and clay
Mor fibrous, poorly decomposed, poorly drained acid humus
Mull soft, black, crumbly, nutrient-rich humus
Parent material rock from which a soil originates
Ped natural soil structural unit made up of many soil particles
Pedology the study of soils
Podsolisation removal in solution of iron and aluminium through the soil profile by organic acids
Regolith layer of loose material resulting from the weathering of exposed rock
Salinisation movement of salts upward, through a soil profile by capillary action
Sesquioxides chemical compounds of oxygen plus a metal
Soil the medium of plant growth, produced by a combination of weathering and the activity of organisms
(soil) Profile variations that occur in the characteristics of a soil along a vertical transect from the surface to the parent material
(soil) Structure way in which individual particles of soil aggregate
(soil) Texture composition of soil in terms of the properties of different-sized mineral particles, i.e. clay, silt and mud

1 Introduction

Together with air and water, **soil** is critical in sustaining life on earth. Soil is formed as part of the general cycle of erosion; the result of rock breakdown through the processes of weathering and erosion, and because soil is part of this natural recycling process, it is also subject to the effects of erosion, transportation and deposition. There are four main components of any soil: mineral materials (which account for about 40% of the soil by volume), organic materials (about 10%) and air and water, which vary inversely with each other and account for a further 45% of the total volume. The remaining 5% is occupied by soil biota. The study of soil is known as **pedology**.

2 Soil Formation

All soils originate from rock, which is known as the **parent material**. This parent material may lie directly below the soil, or may be a great distance away because the soil has been transported by wind, running water or ice. Ice played an important role in the formation of soils across much of northern Europe and North America following the last glacial period. As ice retreated, it deposited thick layers of moraine across many areas of the landscape, and this has formed the basis of later soils. Running water and wind often transport material over long distances – material which, when deposited, affects the subsequent soil development. Material deposited by rivers is referred to as **alluvium** (or as **lacustrine material** if it accumulates in lakes), while material deposited by wind is known as **loess**.

The process of soil formation (**pedogenesis**) is significantly influenced by parent material, climate and biotic activity; to a lesser extent, it is also affected by topography and by the passage of time.

- *Parent material.* Initial soil formation begins with the weathering of exposed rock at the earth's surface or by the deposition of material transported by wind, water or ice. This produces a layer of loose material that is known as **regolith**. A significant proportion of the mineral content (and hence the nutrient content) of any soil is determined by its parent material because it is the product of the rock's chemical composition combined with its rate of weathering; different rock minerals weather at different rates and in differing ways. Similarly, the parent material plays a significant role in determining the depth, **texture** and drainage of the soil. In some cases, particularly in areas of limestone outcrops, the parent material actually determines the type of soil that develops.
- *Climate.* The distribution of soils at the global scale (see Figure 3.1) displays a high degree of correlation with both the world climate map (Figure 2.1) and the world biome map (Figure 2.2). In general terms, climate is a determining factor in the initial breakdown

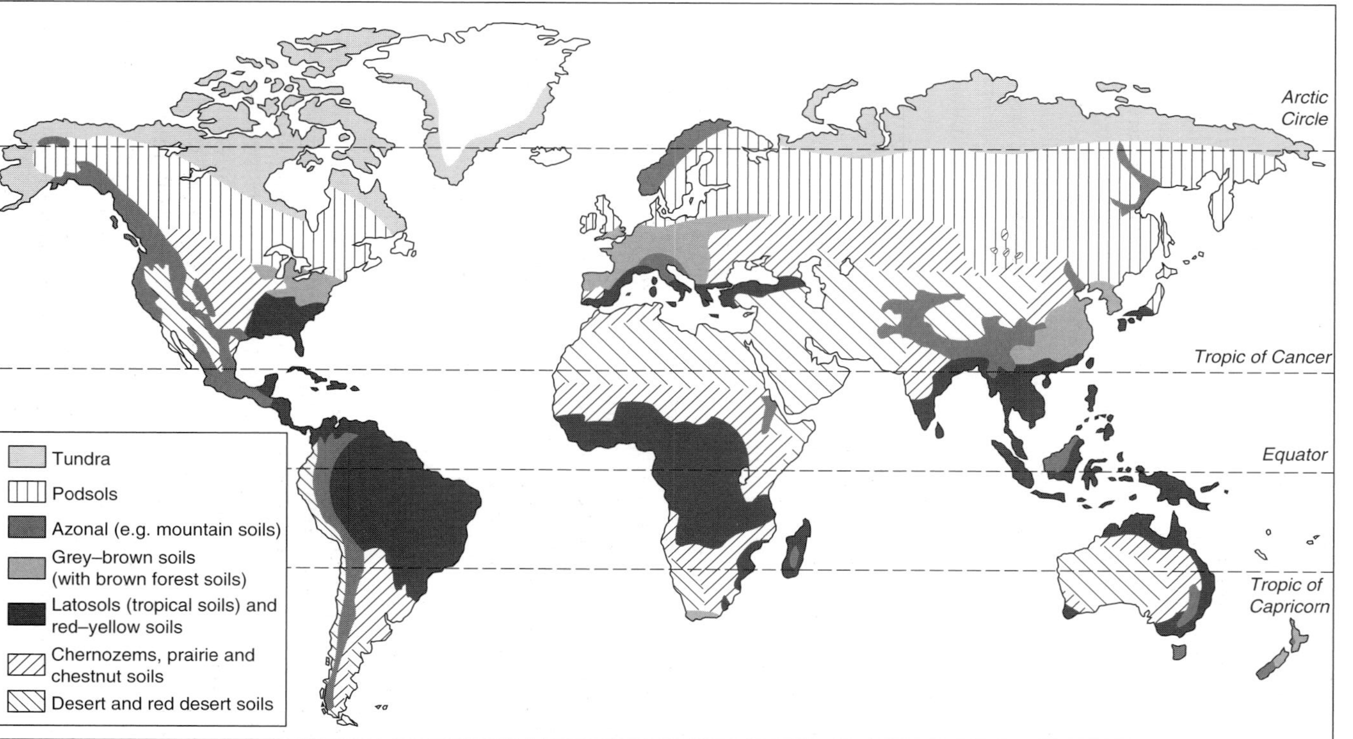

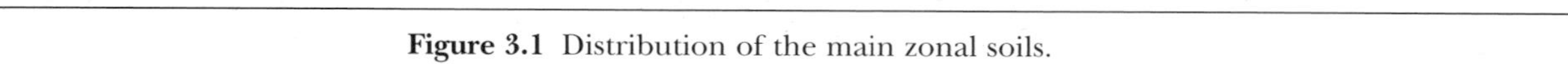

Figure 3.1 Distribution of the main zonal soils.

of rock, with weathering being most rapid at the equator and slowest toward the poles. Both precipitation and temperature influence vegetation growth and, in turn, the supply of **humus** to the soil. In equatorial areas, the supply of humus is greatest, partly because of the volume of biomass available and partly because of the duration of the growing season (see Chapter 4). Towards the poles, vegetation becomes less dense, generally less tall and, in some areas, less woody; this reduces the biomass available for humus production and the growing season becomes shorter, decreasing the input of organic material. Precipitation is also important in its own right. Where rainfall is heavy, **leaching** (see page 44) may occur. Where precipitation is less and evapotranspiration is greater, upward soil movement known as **capillary action** may affect the mineral composition of the soil. Where both temperature and precipitation are greater, decay and decomposition of organic materials are speeded up because high temperatures and increased humidity facilitate the processes of the micro-organisms (soil biota) that are responsible for the detrital processes.

- ***Biotic activity***. The conversion of dead organic material into useful soil nutrients requires the action of a myriad of soil biota. These are often referred to collectively as **micro-organisms**, but more rightly should be recognised as micro-organisms (such as bacteria and fungi) and **macro-organisms** (such as worms, termites and burrowing animals). Not only are these organisms important in the recycling process, they also play a significant role in both mixing and aerating the soil.
- *Topography*. The relief of the land and the local weather are interlinked; as altitude increases, so too does rainfall, cloud cover and the strength (and frequency) of the wind. At the same time, temperatures drop and the length of the growing season diminishes. Where the land is steep, run-off will be greatest and the potential for soil erosion will consequently increase. Soils on steeper slopes are frequently drier and also thinner as a result of this. Where slopes are gentler, the risk of erosion is reduced but soils may become **water-logged** and increasing amounts of water can increase weathering of the underlying parent material. Deposition of soils transported from the upper slopes may occur and it is common to find deeper soils on lower, gentler slopes. Variations such as these often lead to the development of different types of soil at different points across to hill/valley transects; where soil types vary in association with relief, the resultant distribution is referred to as a **catena** (see Figure 3.2).
- *Time*. Soils take a long time to develop to maturity; it takes, on average, between 300 and 1000 years to produce 1 cm of soil. This means that it could take as long as 12,000 years to produce a soil of sufficient depth to support agricultural activity. In hot, humid areas, the time-scale will be shortest, because such conditions are

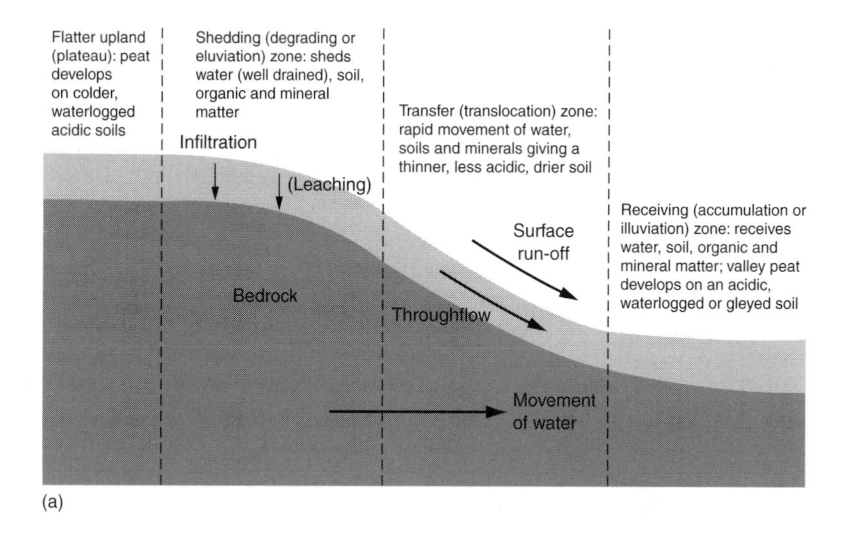

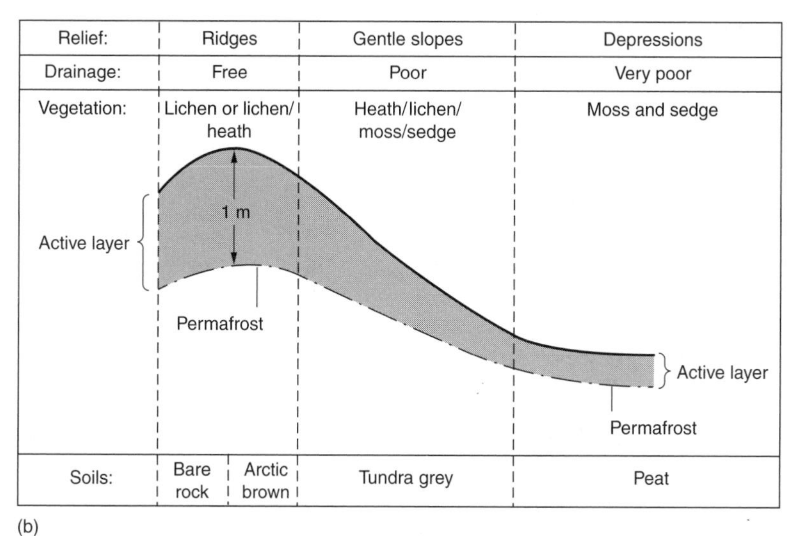

Figure 3.2 (a) Generalised soil catena. (b) Sequence of soil, vegetation and permafrost along a tundra catena.

the optimum for soil development; in cooler climates the process can be far slower. We know that in Britain, soil development has taken about 10,000 years, this being the time since ice melt at the end of the last glacial period. While soils are 'young' they tend to closely resemble their parent materials but, as soil formation proceeds, they acquire new characteristics as a result of organic input, soil biota activities and soil processes (see Figure 3.3). As soils

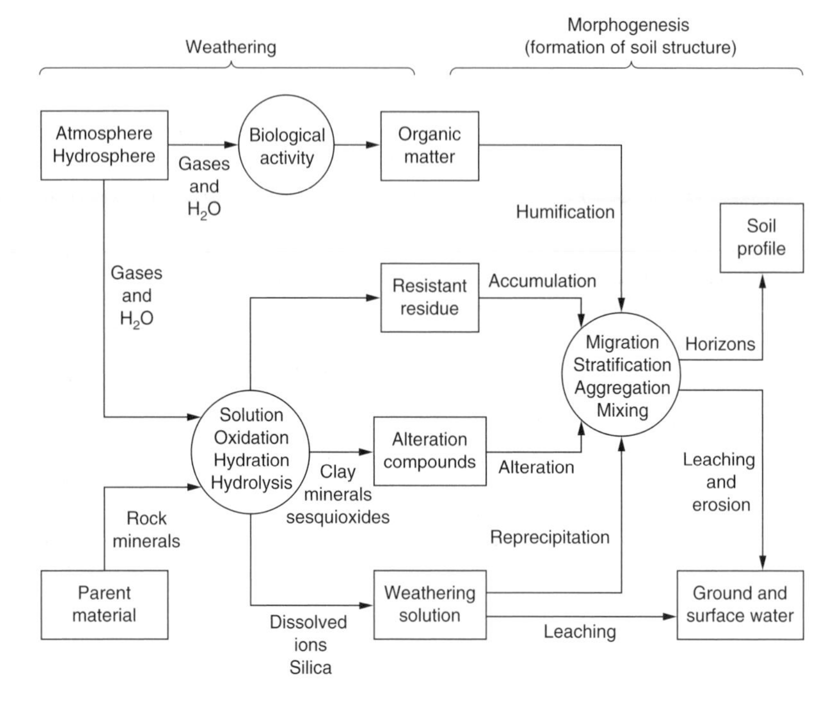

Figure 3.3 Flow chart of soil formation.

mature, they also develop characteristic layering; such strata are referred to as **horizons**.

3 Soil Profile

The vertical, sectional view through a soil from the ground surface toward the underlying rock is known as the **soil profile**. Within a soil profile, the soil horizons can usually be easily identified. Horizons develop over time as the result of the ongoing interaction of the soil with the parent material, the climate and the soil biota. **Immature soils** exhibit few variations through their shallow depth, but with increased weathering, organic input and biotic action combined with the effects of various soil processes (see Section 5 below), clear and distinctive layers begin to develop at different levels within the soil. Immature soil profiles may be less than 10 cm deep, while mature soil may attain depths of several metres. In Britain, most soils are about 1 m deep, reflecting their age of less than 10,000 years. The soil profile and its horizons are the key to understanding the processes that have contributed to the development of a particular soil; they are also the basis of soil classification.

a) Soil horizons

Most soils have three or more horizons and these form the basis for describing and understanding the soil; they are represented in Figure 3.4(a). The upper, **'A' horizon** is often referred to as the 'top soil' or the 'ploughed layer'. It is the zone of concentrated biotic activity and of greatest humus content; therefore it is the source of most plant nutrients and contains the greatest concentration of rooting systems. It is from this layer that materials are leached or **eluviated. Horizon 'B'** is frequently referred to as the 'zone of accumulation' as it is the layer in which leached materials are redeposited or **illuviated**. This stratum is the horizon frequently used to identify and/or classify soil type. **Translocated materials** do not merely accumulate within the 'B' horizon, they generally create sub-layering and consequently the 'B' horizon is often sub-divided in recognition of specific groupings of materials within it. Below the 'B' horizon is the **'C' horizon**. Generally this layer represents the regolith of recently weathered parent material but in some cases, it may represent the deposited layer of materials which have been transported by the agents of erosion and which now form the basis of the overlying soil.

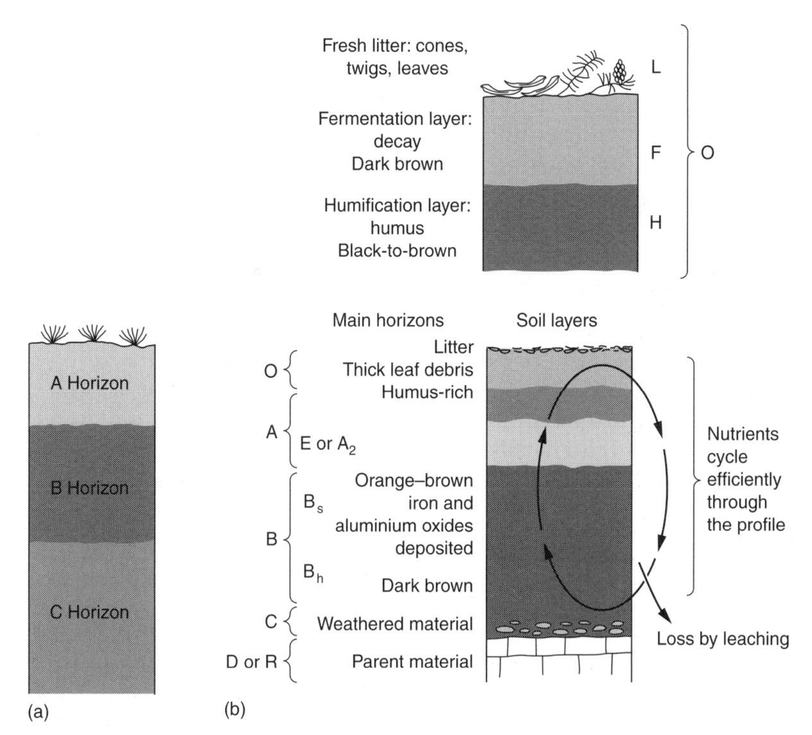

Figure 3.4 (a) Soil horizons and (b) enhanced soil horizons.

In many soils, a simple division into three layers may be over-simplistic (although likewise, some soils have only an 'A' horizon sitting above the 'C' horizon). In recognition of increased complexity, additional sub-divisions both above and within the soil are also recognised, see Figure 3.4(b). Above the 'A' horizon, a distinctive **surface horizon** (referred to by the letter 'O') is frequently discernable; this layer may be sub-divided to recognise:

- a layer of **leaf litter** in which recognisable plant and soil biota can be seen; this is the **'L' horizon**
- a **fermentation layer** (the **'F' horizon**) in which decomposition occurs
- a humified layer (or humus layer), indicated as the **'H' horizon**; within this material, decomposition has proceeded to such a degree that little recognisable plant structure remains in the material. In some soils, humus is well integrated into the upper soil horizon giving it a dark colour, crumbly structure and high nutrient content. In such cases, the humus is referred to as a **mull.** If organic material is slow to decompose (for example in colder climates), it produces an acidic, nutrient-deficient humus layer known as **mor**.

Between the 'A' and 'B' horizons there is often an additional stratum of highly leached materials that is referred to as the 'E' horizon (or the A2 horizon). This is generally easy to recognise as it significantly paler in colour than the horizons above and beneath it. The 'B' horizon sub-divisions indicate the translocation of iron and/or **sesquioxides** ('Bs' horizon) or humus ('Bh' horizon). Any unaltered bedrock may be referred to as either the 'R' or the 'D' horizon.

4 Properties of Soil

Soils are generally porous, open bodies that have the capacity to retain some moisture. Between 40 and 60% of soil's volume is inter-particle space, known as **pore space**. Pores may contain air, or water, or both. An inverse relationship exists between soil air and soil water. Soil water reacts chemically with soil solids and usually contains both dissolved substances and suspended particles.

a) Physical properties

i) Texture

The size of the particles that make up any soil determines many of the properties of that soil. Particles larger than 2 mm are generally disregarded in this consideration. Particles less than 2 mm are sub-divided into three categories:

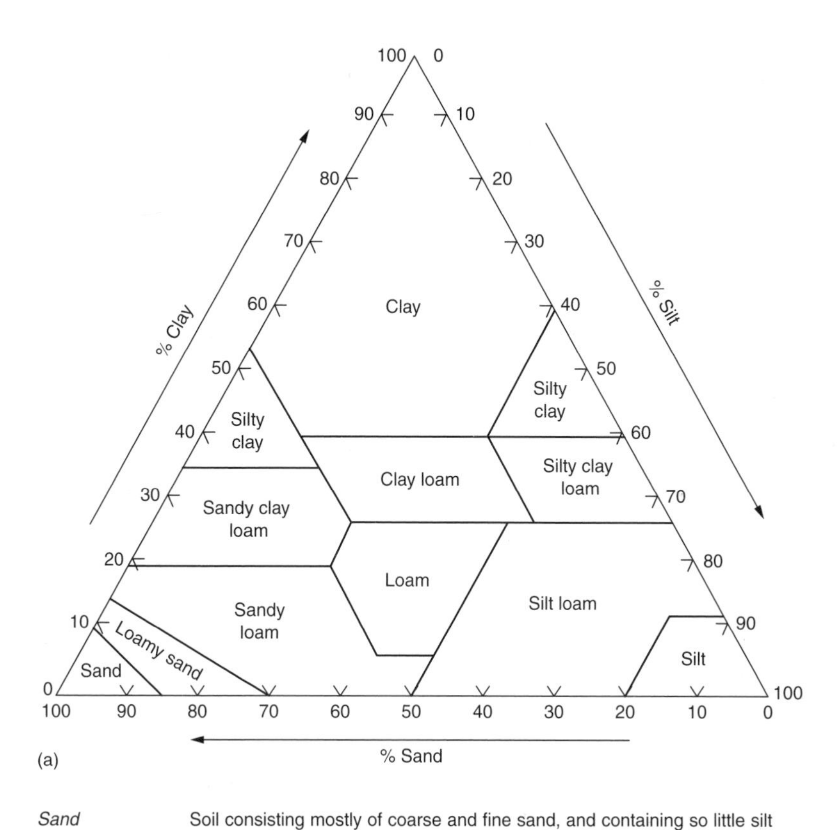

Sand	Soil consisting mostly of coarse and fine sand, and containing so little silt and clay that it is loose when dry and not sticky when wet.
Loamy sand	Soil consisting mostly of sand but with sufficient clay and silt to give slight plasticity and cohesion when very moist.
Sandy loam	Soil in which the sand fraction is still obvious, which can be moulded readily when sufficiently moist.
Loam	Soil in which the fractions are so blended that it moulds readily when sufficiently moist.
Silt loam	Soil that is moderately plastic without being very sticky and in which the smooth soapy feel of the silt is dominant.
Sandy clay loam	Soil containing sufficient clay to be distinctly sticky when moist, but in which the sand fraction is still an obvious feature.
Clay loam	The soil is distinctly sticky when sufficiently moist and the presence of sand fractions can only be detected with care.
Silty clay loam	Soil that contains very small amounts of sand. It is less sticky than silty clay or clay loam.
Silt	Soil in which the smooth soapy feel of silt is dominant.
Sandy clay	The soil is plastic and sticky when moistened sufficiently but the sand fraction is still an obvious feature. Clay and sand are dominant and the intermediate grades of silt and fine sand are less apparent.
Silty clay	Soil which is composed almost entirely of very fine material.
Clay	The soil is plastic and sticky when moistened sufficiently. When moist, the soil can be moulded into any shape and takes clear fingerprints.

(b)

Figure 3.5 (a) Soil textures and (b) soil texture class descriptions.

- **sand** is comprised of particles that are 0.05–2 mm in diameter
- **silt** is comprised of particles 0.05–0.002 mm in diameter
- **clay** is comprised of particles that are less than 0.002 mm in diameter (colloidal clays are particles less than 0.0002 mm which are usually held in suspension in soil water).

Soils are usually a mixture of all three types of particles occurring in different proportions; this provides an effective way of classifying soils, i.e. according to their sand/silt/clay composition. Twelve different classes of soil are recognised and are represented in the triangular graph in Figure 3.5(a) and described in the table in Figure 3.5(b).

Soil texture affects soil air and water content and water movement through the soil because it determines pore space and affects soil structure. Clay soils hold much greater amounts of water but are therefore poorly aerated; however, they are richer in nutrients and organic materials. Sandy soils hold far less water but are well aerated. They are able to retain fewer nutrients, as these are rapidly leached owing to the easier passage of water through them. The 'ideal' soil is **loam**; loam is characterised by comprising all three components in roughly equal amounts (i.e. about 30% each of sand, silt and clay). This provides such soils with an excellent mix of smaller and larger particles, facilitating water retention, drainage and movement. It provides root networks with easy passage through the growth medium and also contains a good supply of air. Soil texture is a relatively stable characteristic that mature soils will retain through time.

ii) Structure

Sand and clay particles behave in very different ways. Clay particles are cohesive (i.e. they stick together), while sand particles lack cohesion and tend to retain their individuality. In the presence of clay, the **primary particles** (i.e. the sand, clay and silt) do not remain in isolation but instead tend to 'stick' together – or are aggregated. Individual particles **aggregate** in a variety of different shapes or forms called **peds**; such aggregation creates the **soil structure**. Both sandy soils and clay soils are described as being **structureless**; sandy soils because the particles lack cohesiveness and clay soils because the particles clump together in **massive units**. Between these two extremes are a range of possibilities (see Figure 3.6). Surface soils are often granular (i.e. they have small peds which are well aerated and drained) while sub-soils may be blocky (i.e. composed of closely fitting peds but usually well drained) or platy (horizontal, elongated peds which restrict the vertical movement of water).

Where soil particles tend to be fine, aggregation increases pore spaces and hence porosity, assisting root growth/penetration and also soil air and water movement.

Most soils are saturated and dry out fairly regularly; many are frozen and then thawed and all are subjected to root and biotic

Ped (type of structure)	Size of structure (mm)	Description of peds	Shape of peds	Location (horizon: texture)	Agricultural value
Crumb	1–5	Small individual particles similar to breadcrumbs; porous		A horizon: loam soil	The most productive. Well aerated and drained – good for roots
Granular	1–5	Small individual particles; usually non-porous		A horizon: clay soil	Fairly productive. Problems with drainage and aeration
Platy	1–10	Vertical axis much shorter than horizontal, like overlapping plates; restrict flow of water		B horizon: silts and clays, or when compacted by farm tractors	The least productive. Hinders water and air movement. Restricts roots
Blocky	10–75	Irregular shape with horizontal and vertical axes about equal; may be rounded or angular but closely fitting		B horizon: clay-loam soils	Productive. Usually well drained and aerated
Prismatic	20–100	Vertical axis much larger than horizontal; angular caps and sides to columns		B and C horizons: often limestones or clays	Usually quite productive. Formed by wetting and drying. Adequate water movement and root development
Columnar	20–100	Vertical axis much larger than horizontal; rounded caps and sides to columns		B and C horizons: alkaline and desert soils	Quite productive (if water available)

Figure 3.6 Types of peds.

activity. The soil movements associated with such events combine to regularly rearrange the individual particles in the soil, bringing particles together as aggregates and dispersing existing peds. While organic materials in the soil reinforce aggregate bonds and so increases their strength, poor soil management or exhaustion of the nutrient content leads to the loss of aggregates and diminishes the soil structure. Unlike soil texture, soil structure is not stable over time.

iii) Soil water

Soil water generally originates as localised rainfall or as overland flow; the amount of water entering any soil is determined by the soil structure. Drainage is the product of the rate of infiltration coupled with the **water retention capacity** of a particular soil. Clayey soils retain moisture for longer but, because clay particles are so small and because such soils are often structureless, infiltration may be reduced. Sandy soils, on the other hand, permit rapid infiltration but have low retention capacities, and when infiltration ceases, much water will drain away in response to gravity. Such water is called **free water** and is quickly lost to the local vegetation cover. Following water input to the soil system and subsequent drainage, the moisture that remains within the soil is referred to as the soil's **field capacity**. Soils with higher clay content tend to retain their soil nutrients (which are held in solution in soil moisture) more effectively than do sandy soils that tend to lose nutrients through leaching.

Soil moisture can be held in one of two ways. Some water is held within the pore spaces while the remainder is held as **hydroscopic water** (i.e. as thin films of moisture around the peds to which it 'bonds'). Such water is not available to plants for water uptake because of the strength of the surface tension fixing it to the soil particles. The water that is available for plant usage is known as **capillary water**; it too is held in the soil by cohesion – this time cohesion to the hydroscopic water, but the cohesion in this case is very loose and can easily be broken when water is needed for root absorption.

During periods when evapotranspiration exceeds precipitation, plants may utilise all of the available soil moisture. When this happens, the vegetation is said to experience **water stress**. If soil moisture is not replenished at this point, the plants will quickly reach their '**wilt point**' and their ability to photosynthesise will diminish. Plant death will follow if the plants are not xerophytic (i.e. adapted to cope with aridity) or if water does not reach the soil soon.

iv) Soil air (atmosphere)

Soil air is a mixture of gases occupying the portion of pore spaces that are not filled with water. It is sourced from the atmosphere but has slightly different characteristics. Primarily it contains less oxygen but more carbon dioxide – up to 10 or 100 times more carbon dioxide. These differences are mainly the result of the respiration of the soil biota. If soils become waterlogged, the affected soil is described as being **anaerobic**. Anaerobic conditions can lead to the death of soil organisms and root systems, not only because of the lack of oxygen but also because of the build-up of toxins (particularly hydrogen sulphide) under anaerobic conditions.

v) Soil temperatures

Soil temperature is determined primarily by the amount of sunlight that is available to it and the amount of water it contains. Soils are warmed by solar radiation and therefore dark-coloured soils warm most quickly. Vegetation cover plays a part in determining the amount of solar radiation that reaches the soil surface. Because the top few centimetres of soil can become very warm, there are usually fewer micro-organisms in the top-most horizon. Temperatures also affect the rates of chemical and biological reactions in the soil. Generally, biotic activity increases as temperatures rise (unless it so hot that the organisms die) but slow (and may be arrested) as temperatures fall towards freezing.

vi) Consistency

Consistency refers to the soil's physical condition at various moisture content levels and is important in relation to the soil's ability to cope with stress (e.g. drought) and manipulation. The consistency is determined by a combination of soil texture, organic content (which affects moisture retention) and the type of clay minerals present in the soil.

vii) Colour

The colour of the soil often reflects its moisture content and varies with that content. Darker soils contain more water and organic material. Usually the 'A' horizon is darker than those beneath as it has the highest levels of organic material. Red, yellow and grey colourings indicate the oxidation/hydration states of any iron oxides that are present. Red or yellow colours suggest the presence of the good drainage and aeration necessary for the soil biota to function well. Grey soils are generally poorly aerated and often have a very high soil moisture content.

b) Chemical properties

Eight chemical elements make up the majority of the mineral matter in the soil. Oxygen is the most prevalent; oxygen is a negatively charged ion (**anion**). The remaining seven minerals are all positively charged ions (**cations**) and, in descending order of importance, are: silicon, aluminium, iron, magnesium, calcium, sodium and potassium. Over 80 other chemical elements are also found in varying proportions, but very small quantities. Mature soil is significantly different chemically from both the parent material and the regolith. It contains less calcium, magnesium, sodium and potassium (i.e. the water-soluble minerals) and more of the insoluble minerals of iron and aluminium. Very old, well-weathered soils (such as those of the tropical rainforests) often have very high proportions of iron and aluminium.

Organic matter in soils has an important influence on the soil's chemical properties. The organic matter is composed mainly of carbon, hydrogen, oxygen and nitrogen, together with traces of sulphur and other elements. Organic materials increase soil holding capacity and cation exchange capacity (see below) and also improve soil aggregation and structure. Together with colloidal clays, they are the most active of all the soil constituents.

i) Cation exchange

Silicate clays and organic material are usually negatively charged and therefore attract positively charged cations to them. **Cation exchange** describes the ability of soil particles (clay and organic matter) to **adsorb** (i.e. to hold molecules of a gas, a liquid or a solute on the outer surfaces of each particle) and exchange cations with those held in **soil solution** (i.e. in water held in pore spaces). Cation exchange capacity is important because:

- it allows cations of calcium, magnesium and potassium to detach from clay and humus particles and become available for plant uptake
- cations adsorbed to exchange sites are more resistant to leaching
- hydrogen is released by the process and consequently increases the acidity of the soil
- increased acidity subsequently increases the weathering of the parent material which in turn releases more minerals which can be utilised for plant growth.

The cations of calcium, magnesium, potassium and sodium produce alkaline reactions in water and are known as base (or basic) cations (or **bases**). Aluminium and hydrogen ions produce acidic reactions in water and are known as acidic cations. The greater the proportion of base cations (bases), the higher the soil's pH and the more alkaline the soil.

ii) Soil pH

Acidity is one of the easiest aspects of soil to measure and, like taking a person's temperature to assess illness, pH is used as a general measure to tell us much about the body of the soil. Soil pH will increase as acidity decreases; pH 7 is considered to be neutral with readings >7 indicating alkaline conditions and readings of <7 indicating acidity. Generally soil pH varies from about 4 to 8.5, but can be as low as 2. pH is important because it has a direct effect on the quantity and activity of the soil micro- and macro-organisms. This in turn affects decomposition and the recycling of nutrients; which duly affects the vegetation cover and the next input of nutrients for recycling.

Phosphorus is most readily available where soils are either slightly above or below pH 7; all other nutrients become more readily available as soil becomes more acidic but elements such as aluminium and iron become toxic to plants if acidity exceeds 5.5. Bacteria and other micro-organisms are most active when pH is closer to 7 or slightly alkaline; therefore, a neutral or slightly acidic soil provides the optimum conditions for soil function.

c) Biological properties

Soil contains a huge range of life-forms, from viruses, bacteria and algae, termites, woodlice and earthworms to large burrowing mammals such as badgers and rabbits. Microscopic life is known as **soil microflora** (even though some of these life-forms are not strictly plants), while larger beasts are referred to as the **macrofauna**. Most animals improve the soil's physical properties through burrowing and feeding because these activities improve drainage and aeration. They do little to alter the chemical properties of the soil, although by improving physical conditions, they may enhance microbial activity.

The diversity of micro-organisms is enormous in terms of both form and function. The most important groups are bacteria, fungi, algae and protozoa. Because many of these organisms are so very tiny, huge numbers can inhabit a very small volume of soil and space is never an inhibiting factor for them. Soil fungi contribute the greatest biomass to the soil and are usually considered to be the dominant group of decomposers in the soil. Soil microbes interact, like all other parts of an ecosystem, not only with other microbes like themselves but also with other microbe groups. In fact, it has been discovered that decomposition occurs more effectively and quickly where there is a mixed population of microbes rather than only one or two groups. Microbes, considering their tiny size, are at the heart of the nutrient recycling system, controlling soil fertility and plant productivity. They are also critical in producing carbon dioxide, nitrogen, sulphur and phosphorus for plant utilisation. In agricultural areas, the mineralisation of nitrogen by microbes is of critical importance in sustaining high crop yields.

5 Soil Processes

Soil processes describe and highlight the effects of the passage of water through the soil profile and the consequent translocation (movement) of materials which result. It is these processes that significantly contribute to the development of the 'A2' and 'B' horizons. Within the surface 'O' layer, the processes of **humifaction** (the decomposition of organic matter to produce humus) and **cheluviation** (a specialised form of **eluviation** in which the minerals of iron and aluminium are dissolved by the organic acids released during decomposition and then translocated downwards through the soil profile) are both dominant. Within the soil profile itself, any one of six other processes may (or may not) operate, contributing to the development of an 'A2' and/or the 'B' horizon. These are considered below.

a) Leaching

Leaching is the removal of soluble material in solution; it occurs in areas where precipitation exceeds evapotranspiration throughout the year and where soil drainage is effective. Acidic rainwater causes the breakdown of clays in the upper horizons and also dissolves soluble salts; these are then eluviated from the top soil and illuviated in the lower, 'B' horizons where they may create identifiable sub-strata. See Figure 3.7(a).

b) Podsolisation

Podsolisation is the extreme form of leaching; it is most commonly found in cooler biomes where precipitation greatly exceeds the demands of evapotranspiration and where soils are sandy and therefore drain overefficiently. Under coniferous forest particularly (but also in areas of heathland) the acidity of rainwater is enhanced and this dissolves still more bases together with the sesquioxides of iron and aluminium. Redeposition of the iron at the top of the 'B' horizon often creates an impenetrable layer known as **hard pan**. See Figure 3.7(b).

c) Gleying

Gley soils are associated with waterlogging of the soil body and are generally found on lower, gentle relief features. The anaerobic conditions may affect the soils temporarily (e.g. after a flood event) or may be more permanent. Where the soil is underlain by impermeable rock, the process of gleying will be enhanced. Under anaerobic conditions, iron is chemically weathered to become blue–grey ferrous iron (Fe^{2+}). Because gleying is the result of poor, localised drainage, gley (or gleyed) soils may occur world-wide; it tends to be a soil process that affects many soils in Britain to a greater or lesser degree.

d) Ferrallitisation

The process of **ferrallitisation** is predominant within the tropical biomes in areas where the parent material is weathered to produce **kaolinite** (clay) and sesquioxides (hydrated oxides of iron and aluminium). Its name refers to the translocation of iron (Fe) and aluminium (Al) owing to rapid chemical weathering under high temperatures and effective soil moisture. Where rain falls throughout the year, **ferrallitic soils** develop; where drought affects the soil for some part of the year, **ferruginous soils** are created. Ferruginous soils may develop a stratum of **laterite** (hydrated iron and aluminium oxides) close to the soil surface as a result of seasonal variations in soil water movement. See Figure 3.7(c).

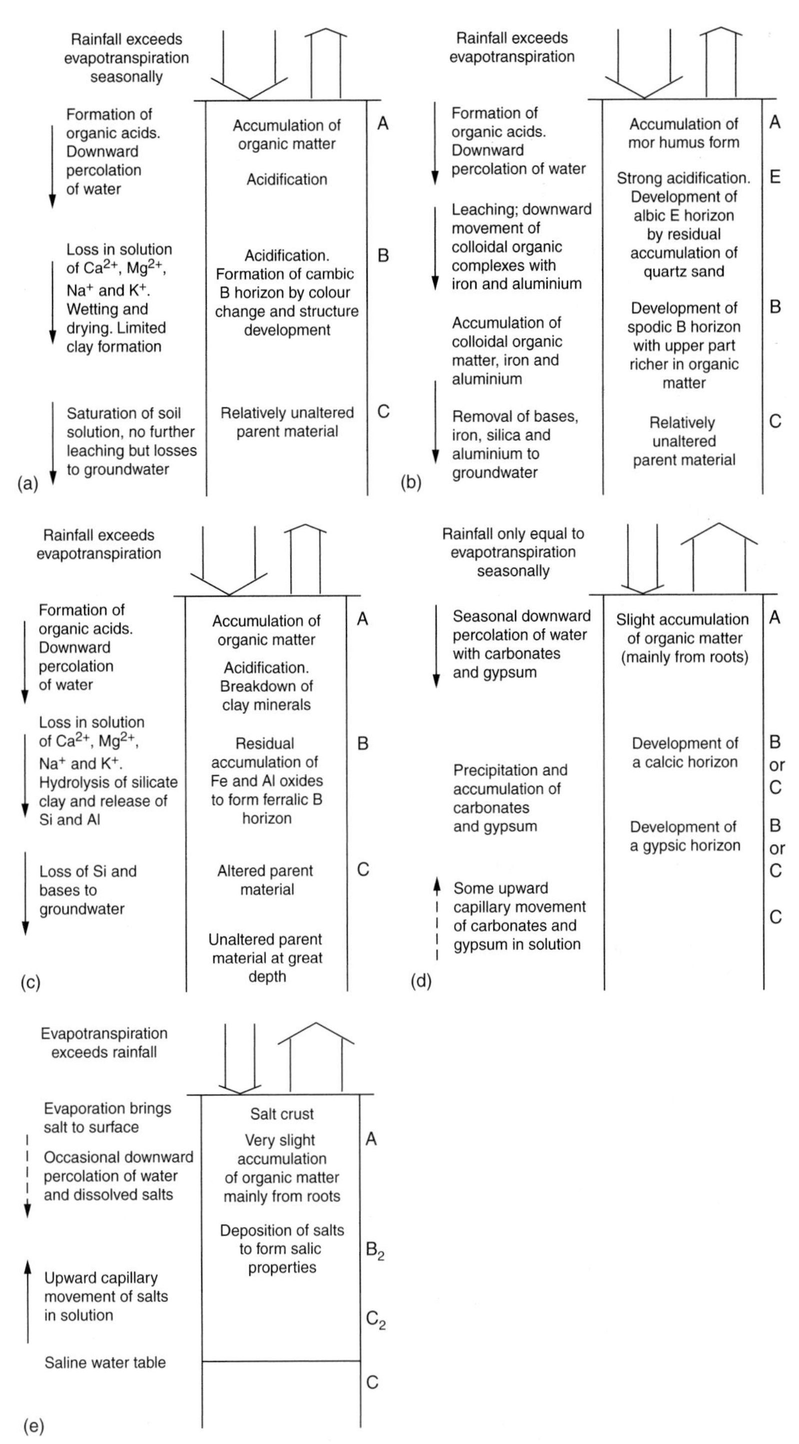

Figure 3.7 The processes of (a) leaching, (b) podsolisation, (c) ferrallitisation, (d) calcification and (e) salinisation.

e) Calcification

Calcification occurs in areas where precipitation roughly equals the evapotranspiration rates. Any leaching which occurs at times when $P > eT$ is mild, and at other seasons, when $P < eT$, capillary action draws water to the surface. Leaching is insufficient to remove completely the calcium bases from the soil, and any which are translocated are redistributed upwards again during periods of capillary movement. Thus the 'B' horizon remains proportionally rich in calcium when compared to its content of other bases and the soil is described as being calcium rich. See Figure 3.7(d).

f) Salinisation

Salinisation of soils is associated with areas where **potential evapotranspiration** greatly exceeds the precipitation and where the water table is shallow. Excessive evaporation leads to the upward movement of water through the soil by capillary action. Salts in solution are drawn upwards with the water and, as pure water is evaporated at the surface, they are redeposited at (or near to) the surface creating a hard, salt 'crust'. Salinisation is an increasing problem in arid areas where irrigation has been introduced to permit cultivation but is poorly managed – with overwatering of the crops creating excess soil moisture and consequent capillary action. See Figure 3.7(e).

6 Soil Types and Classification

Soils are exceptionally difficult to classify; this is partly because soil forms an almost continuous covering of the earth's surface and partly because soil is below ground and therefore cannot be easily accessed for study. Soils are not discrete units; they grade into other soils across the landscape without obvious demarcation at the surface – or even obvious demarcation in profile.

Soil texture offers an obvious and easily measured classification (ranging from clayey through to sandy) that may be applicable at the local level. However, it is not an adequate way of classifying soils at regional or global levels. Too many important variables are excluded from such a simplistic system. There are thousands of different soils when considered at the global scale. This is not surprising, considering the great range of climates, ecosystems and relief occurring across the globe. So far, pedologists have been unable to formulate a universal differentiation system, but two classification systems are in common usage at present:

- the US Soil Conservation Services' 'Soil Taxonomy', which is based on a wide range of soil properties that can be both observed and objectively measured

- the UNESCO/FAO classification, which is similar to the US system but lacks the same degree of rigorous measurement.

All attempts at classification use the concept of the soil profile in combination with characteristics such as organic content, texture, colour and stratification.

However, an alternative way to group soils is to assess their formation to determine either the role played by the interaction of factors such as climate and vegetation cover, or the dominance of a single factor such as bedrock.

Zonal soils are considered to be mature soils – representing the interplay of climate, vegetation and parent material across a significant period of time. They are generally seen as the product of stable conditions in areas that do not experience extremes of weathering, relief or drainage. Because they represent such stability, zonal soils exhibit a relatively deep profile showing distinct horizons. These may be further sub-divided as a result of continued translocation of materials within the soil. **Azonal soils**, in contrast, are immature soils representing a continuing formation process. They remain dominated by the characteristics of their parent material and lack defined horizons. The 'A' horizon may lie directly on the 'C' horizon due to the absence of the action of soil-forming processes. Soils forming on sand dunes, in salt marshes and on newly erupted volcanic materials are included in this type of soil. **Intrazonal soils**, like azonal ones, are not associated with specific biomes or climates. They are soils in which the formation process has been dominated by a single factor such as parent material (e.g. **calcimorphic soils** such as rendzinas in limestone areas) or water-logging (e.g. **hydromorphic soils** such as gley/gleyed soils) and can be found world-wide.

Intrazonal soils such as **rendzinas** commonly form in upland areas such as southern England, whose distinctive natural landscape comprises a series of chalk and limestone ridges separated by clay vales. The underlying geology of the ridges encourages free drainage because both limestone and chalk are highly permeable, but chemical weathering means that they produce few residual materials from which soil might develop. Typically, such soils are thin and unable to support a climax vegetation of woodland; instead, the mature vegetation cover takes the form of densely matted grasslands. As a result of the alkaline parent material, vegetation is rich in bases. This encourages the soil macro- and microfauna to recycle material rapidly. The resulting humus is a rich and darkly coloured mull, producing an 'A' horizon so dark as to appear black in colour. This horizon is invariably thin, and often lies directly over the 'C' horizon. The lack of a 'B' horizon is due to the clay content of the upper stratum of soil, which prevents leaching (see Figure 3.8). Rendzinas are alkaline soils with a pH value of 7–8, the result of calcification and a lack of hydrogen ions.

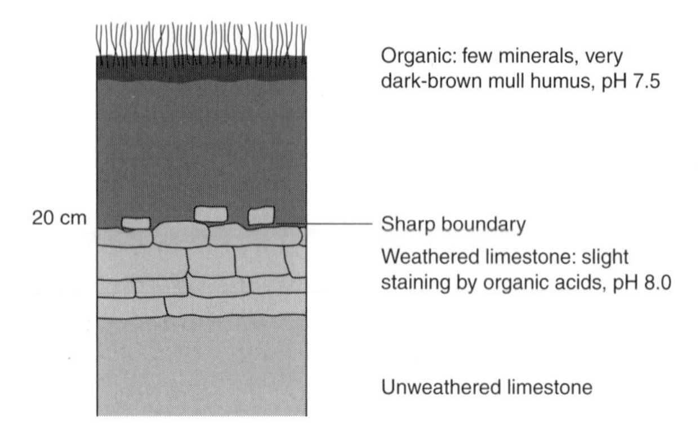

Figure 3.8 A typical rendzina soil profile.

7 Soil Function

The most obvious function of soil is as a growth medium for crops and other plants. Soil also supports the practice of silviculture (the cultivation of trees for harvest) and maintains the natural vegetation cover of land-based ecosystems and biomes. It also plays a significant role in supporting the earth's **fauna**; it hosts a myriad of micro-organisms (up to 20,000 species of bacteria alone may live in 1 gram of soil), and in supporting the natural vegetation of the earth's ecosystems, soil also provides food and homes for the wealth of wildlife and domesticated animals.

Recently, scientists have begun to recognise the important role that soil plays in maintaining the quality and distribution of water supply. The nature of topsoil influences whether rainwater will infiltrate the soil (to become part of the soil store or the groundwater store) or become run-off, joining stream, rivers or reservoirs.

Soils also play an important role in pollution control – especially the control of nitrates and pesticides. The proportion of organic material, the structure and the texture of soil all affect the proportion of fertiliser or pesticide which is retained in the soil and how much subsequently reaches water bodies (including aquifers) and hence the water supply network.

Additionally, soil is an important consideration for civil engineers engaged in major construction programmes. It forms the basis of the foundations for buildings, roads and bridges and ultimately, soil structure and texture will therefore determine the stability of completed structures.

Finally, soil has an important part to play in the climate change equation. The organic material in soil is a major carbon store and is therefore both a major driver and responding variable to change; it may function as a carbon sink or as a carbon source. Additionally, soil may serve as a sink or a source of both nitrous oxides and methane – themselves major contributors to the greenhouse gas effect.

8 Soil Preservation

Like water and air, soil supports life on earth. The current projection of a world population of eight billion by the year 2025 highlights the need to conserve the earth's soil cover. However, for much of the twentieth century, there was little appreciation of the importance of soil conservation; in many areas soil was damaged beyond repair or, in other cases, catastrophe was narrowly avoided only by last-minute interventions.

Soil conservation and desertification are major problems in every continent and not only in drier (drought affected) countries (see Chapter 7). Over 25 million ha of soil are lost to soil erosion every year. Even though erosion is a natural process, this rate is excessive and is mainly the result of poor management practices that leave soil surfaces exposed to wind and running water. UNESCO, FAO, the EU and US conservation bodies have all taken steps to highlight the problems of soil erosion and reduce its occurrence, but the problem continues to worsen. With even more people to feed, food production must increase and therefore we cannot afford to lose more productive soil; but the likely impact of global warming may make the struggle to control soil erosion yet more difficult in the years to come.

Salinisation (and alkalisation) affect over 20 million ha of land, mostly in arid and semi-arid regions of the world. Some salinisation is the result of in-blown salt deposition but most is the result of ill-managed irrigation schemes which overwater crops and so increase the capillary movement of soil water owing to high evaporation rates. Once salt occurs in soil it is both difficult and expensive to flush away.

Intensive cultivation is one of the results of global population increases during the twentieth century. Intensive agriculture has resulted in deforestation and the clearance of land for cultivation coupled with the use of heavy machinery and has led to:

- soil compaction
- damage to other physical properties of the soil
- loss of organic material
- overuse of fertilisers and pesticides, which has led in turn to contamination of water supplies and loss of soil biodiversity.

Added together, the result of such damage is a decline in soil 'health' and increasing fragility of the soil body.

Continuing urbanisation and industrialisation means that increasing amounts of land are being lost; in less economically developed countries (LEDCs) over 1% of the agricultural soils are lost every 10 years to urban developments. Industrialisation leads to acid rain (see page 146), which in turn increases soil acidity; some areas are now recording soil pH levels of 4 or less. This also reduces soil well-being for, as noted above, optimum soil pH is in the region of 6.5–7.

Today, many soils around the world are damaged and increasingly fragile. Even arresting such damage is costly and difficult. Reversing it is unlikely. In many LEDCs, the farmers' primary concerns are to feed the family. Even management practices as simple as fallowing cannot be contemplated when there is no other source of food. The causes of these problems are well documented, but initiating successful remedial practices has not often proved successful; even in the 'Dust Bowl', government incentives to increase production during the 1970s caused farmers to remove wind breaks to maximise cropping areas, only to see their soil eroding again within a decade through the return of drought and strong winds. Successful programmes will necessitate protracted and close co-operation between intergovernmental agencies, non-governmental bodies and local farmers. Soil takes thousands of years to develop, but takes very few years to degenerate.

Summary

- The study of soil is known as pedology, and the process of soil formation is called pedogenesis.
- Soil is formed as a result of the weathering and erosion of surface rocks and is itself subject to the processes of erosion, transportation and deposition. The rock from which a soil is derived is known as its parent material. This weathers initially to form a layer of loose rock, referred to as the regolith.
- There is a strong, positive correlation at the global scale between soil type and its major determinant, climate.
- The vertical sectional view through soil from the surface to the underlying rock is known as the profile. Eight soil processes contribute to the development of the soil profile: calcification, cheluviation, ferrallitisation, gleying, humifaction, leaching, podsolisation and salinisation. As soils mature, they develop characteristic layers (strata) known as horizons.
- Eluviation (leaching) refers to the downward movement of materials (often sesquioxides and silica) from upper strata to lower ones. Materials moved in this way are said to have been translocated. **Illuviation** is the process by which leached materials are redeposited within the lower horizons of a soil.
- A soil's texture is determined by the size of the individual particles and this affects air and water content as well as the movement of moisture through the soil. The 'ideal' soil is loam – a mixture of sand, silt and clay in roughly equal parts.

- Soil structure refers to the aggregation of its particles. Individual units of soil are called peds and these may exhibit a variety of shapes. Structure affects water retention, drainage and root penetration.
- The organic content of soil is very important because it increases the soil's moisture holding capacity and its cation exchange capacity, and improves both its aggregation and structure. Microbes are at the heart of the soil's nutrient recycling capability; they also control soil fertility and, therefore, plant productivity.
- Soils are very difficult to classify, partly because there are so many global variations.
- Zonal soils are mature soils that are the result of stable conditions over a long period of time; they tend to exhibit relatively deep profiles with clearly recognisable profiles. Azonal soils are immature soils that lack horizons and are still in the process of developing; 'soils' in sand dune and salt marsh areas are considered to be immature. Intrazonal soils are the product of a single causal factor, such as parent material or waterlogging.
- The major function of soil is to act as a growth medium for natural vegetation and crops. It is, therefore, crucial to the continuing existence of the earth's terrestrial animal species, and soil erosion, degradation and conservation are currently global issues of particular concern.

Student Activities

1. a) Explain what you understand by the term 'leaching'.
b) What is the difference between 'eluviation' and 'illuviation'?
c) Describe what is meant by the process of podsolisation, illustrating your answer with a sketch diagram.
d) Explain how a parent rock of sandstone might further increase the acidity of a podsolic soil.

2. To what extent do climate and geology affect soil formation?

3. a) What do you understand by the term 'soil'?
b) Explain the process of 'pedogenesis'.
c) Discuss the ways in which human activity can affect the characteristics of soils.

4 Tropical Biomes

KEY WORDS

Carbon sink area in which the vegetation's intake of carbon dioxide exceeds its output of this gas
Continentality extent to which the climate of a place is influenced by land-mass size and distance from the ocean
Deforestation deliberate clearance of forest by cutting or burning
Desert arid area with a total annual precipitation less than 250 mm
Desertification degradation of formerly productive land, caused primarily by human activity and climatic variation
Halophytic plant adaptation to saline conditions
Inter-Tropical Convergence Zone (ITCZ) zone of low atmospheric pressure where the north-east and south-east trade winds meet
Laterite hardened layer of ironstone within the soils of tropical areas
Mycorrhizal association symbiotic relationship between fungi and tree root systems
Pyrophytic plant adaptation to survive natural fire events
Resource partitioning dividing up of scarce resources so that species with similar requirements use them at different times and in different ways and places
Savanna tropical grassland occurring between tropical rainforest and hot desert
(tropical) Rainforest climax vegetation type associated with equatorial regions and/or climates; within the Amazon Basin this is known as **selva**
Xerophytic plant adaptation to survive drought conditions

1 Introduction

Within the tropical areas of the world, precipitation is the chief determinant of vegetation and, hence, ecosystem development. Plants require warmth, light and moisture in order to thrive and, within the tropical and sub-tropical areas, both warmth and light are available in abundance. Equatorial areas provide optimum growing conditions but with increasing distance from the Equator, precipitation totals and reliability both dwindle and, in response, biomass diminishes likewise.

Tropical areas are increasingly a focus of environmental, ecological and humanitarian concern. The continuing destruction of the **rainforests** highlights issues of global warming and **desertification**. At the same time, increasingly unreliable rainfall coupled with unsustainable farming practices across large areas of **savanna** has accelerated the desertification process and led to the spectre of mass famine, not least across sub-Saharan Africa.

2 Tropical Rainforests

Tropical rainforests are an example of broadleaf, evergreen forest and play a crucial role in the oxygen/carbon dioxide cycle that provides the oxygen essential to most life. Undisturbed, they occupy a broad band within the equatorial regions of the world in which the constant 'greenhouse' conditions are ideal for plant growth. Unlike most other biomes, soil nutrients – not water and temperature – are the limiting factor to vegetation growth.

It is generally believed that equatorial regions have experienced little or no climatic change for millions of years, making the tropical rainforests the oldest of all earth's ecosystems. This past environmental stability has probably contributed to the rich biodiversity that is characteristic of life in the rainforests, and it is estimated that the Amazon rainforest alone is home to more that 50% of the world's **flora** and fauna. It is also commonly believed that, having been largely undisturbed by past human activity, rainforest vegetation provides an example of a classic mature (climax) community. Recent archaeological evidence, however, suggests the opposite: that previous human populations and interventions have been far greater in the past and that the rainforest of today is actually a *secondary* vegetative succession, i.e. largely the result of human interaction across thousands of years.

a) Location

Generally, the tropical rainforest biome occupies a broad swath between latitudes 10°N and 10°S, although *true* tropical rainforests (those which are the result of double rainfall/temperature maxima) are restricted to 5°N and S. With increasing distance from the equator, the influence of a progressively drier (but not a 'dry') season triggers subtle changes to both flora and fauna. Where rainfall is less than 120 mm for more than a month, tropical forests should be more accurately referred to as **Tropical Moist Forests**; where a distinctly drier season can be recognised, trees become truly deciduous and the resultant forests are known as **Tropical Deciduous Forests**. Interruptions to the zone of tropical rainforest also occur in high altitude areas such as the Andes, where mountain biomes occur instead. As altitude increases, **Montane Forest** develops in response to lower temperatures and heavier cloud cover. Trees in such areas are less tall and less densely clustered, providing the opportunity for greater vegetation cover at ground level – both as **undergrowth** and on the forest floor itself.

Deforestation has significantly reduced the proportion of the earth's land surface covered by tropical rainforest. Its largest single expanse occurs in the Amazon Basin, where it is referred to locally as **selva**; other major areas occupy West and Central Africa (particularly the Congo Basin), Central America, Southeast Asia (especially Malaysia and Indonesia) and the coastal areas of northern Australia (see Figure 2.2).

b) Climate

Typically, the equatorial climate exhibits little change over days, seasons or years. Mean temperatures are uniformly high at about 27°C and vary by only 1° or 2°; they rarely exceed 32°C. Diurnal temperature ranges can vary much more widely, with a typical range of 10–15°C. Annual precipitation is over 2000 mm and may exceed 3000 mm. Rain falls throughout the year – although the generalisation that it falls every day is untrue. On average, rain falls on about 250 days per year (see Figure 4.1a). Precipitation is intense convectional rainfall, often accompanied by thunder. The combination of high rainfall and temperatures produces high humidity – usually about 80% but sometimes rising to 100%, making the environment uncomfortably hot and 'sticky'. Insolation is fairly evenly distributed throughout the year and most equatorial regions experience an even 12/12-hour split between daylight and darkness. A drier season may begin to appear with increasing distance from the Equator, coinciding with the period when the **Inter-Tropical Convergence Zone** (ITCZ) moves away after an equinox. Where such drier seasons do occur, temperatures tend to be slightly higher during the period of reduced rainfall. Closer to the Equator, double maxima of temperature and rainfall occur as both the sun and the ITCZ are directly overhead twice a year. Figure 4.2 shows the significance of the ITCZ migration to changes in the annual distribution of rainfall in 'the tropics'.

c) Vegetation

Tropical rainforests are uniquely diverse in terms of both flora and fauna, providing habitats for 50–80% of terrestrial plants and animal species – even though they cover less than 5% of the earth's surface. Intense competition for sunlight has led to the evolution of specialised plants and animals that occupy specific **niches** within the stratified forest (see Figure 4.3). Such specialisation enables a huge variety of plants and animals to co-exist without undue competition for resources, a process known as **resource partitioning**. Unlike temperate forests, no single vegetation species dominates the forest, nor are there large stands (clusters) of single tree species. It is quite usual for a species' nearest neighbour to be several hundred metres away from it. A single hectare of tropical rainforest is estimated to contain over 200 species of plants and about 40,000 different species of insect.

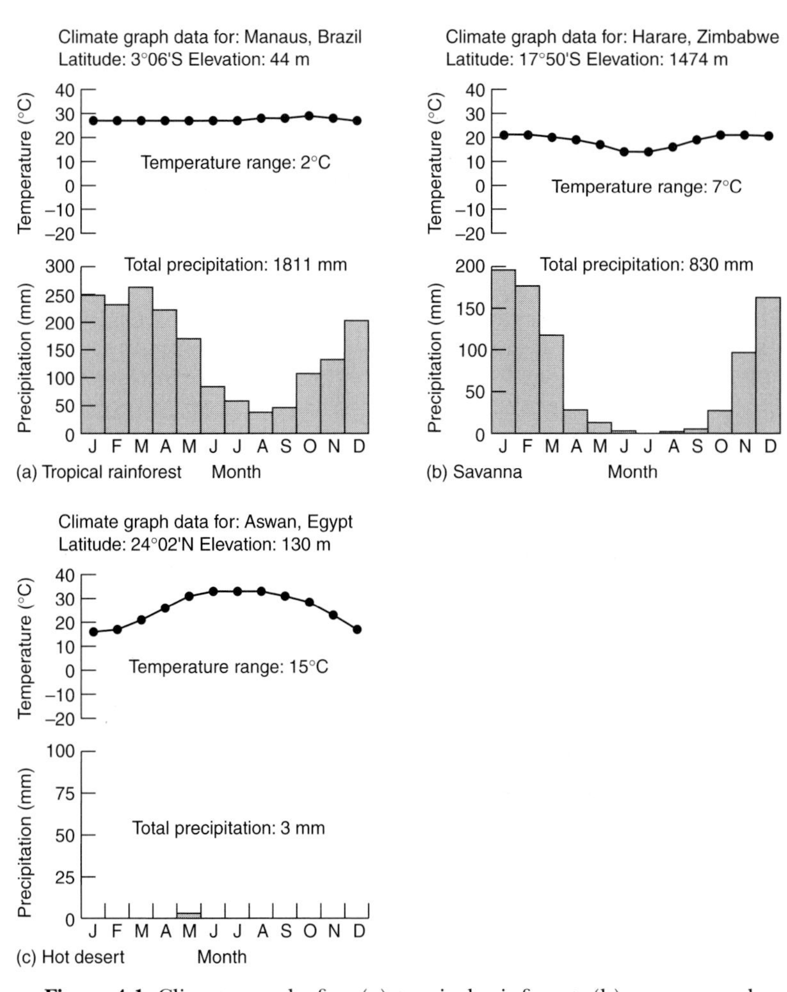

Figure 4.1 Climate graphs for: (a) tropical rainforest, (b) savanna and (c) a hot desert.

While the forest as a whole is evergreen, individual trees are deciduous. In the absence of any cool/cold seasons and no marked dry season, growth is continuous and the deciduous trees are aseasonal (i.e. they shed their leaves throughout the year). Figure 4.3 shows how a typical section of forest consists of five or six strata (layers), each of the three upper strata evolving differently shaped crowns of foliage to maximise sunlight reception at their own particular level.

Important tree species include the world's most valuable hardwoods (teak, mahogany and rosewood), rubber trees, wild banana and cacao (cocoa-producing) trees. A range of rainforest adaptations is also shown in Figure 4.3, but others include:

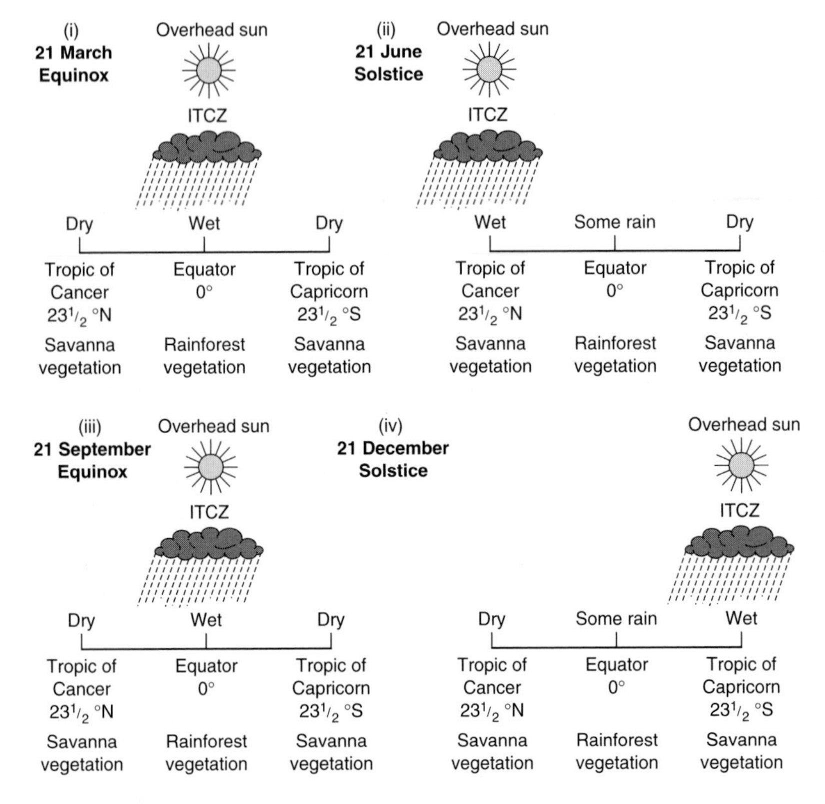

Figure 4.2 The relationship between the movement of the overhead sun, atmospheric pressure and rainfall along N/S transect.
Note: ITCZ (the Inter-Tropical Convergence Zone) is where opposing north-east and south-east trade winds meet.

- Thin, smooth bark, because protection from low temperatures is unnecessary.
- Leathery/waxy leaves that protect the leaves from burning in the intense sunshine.
- Large leaves to encourage transpiration.
- Drip-tips and waxy leaf surfaces to prevent the leaves from becoming 'home' to the large variety of mosses and algae which inhabit the canopy cover. Both features allow the leaves to dry very quickly after rainfall and so become inhospitable to such growths, which would otherwise block their stomata and reduce their ability to photosynthesise.
- Individual leaves that are able to 'twist' or rotate on their stem, allowing each leaf to maximise its exposure to direct sunlight throughout the day and so increase its ability to photosynthesise.

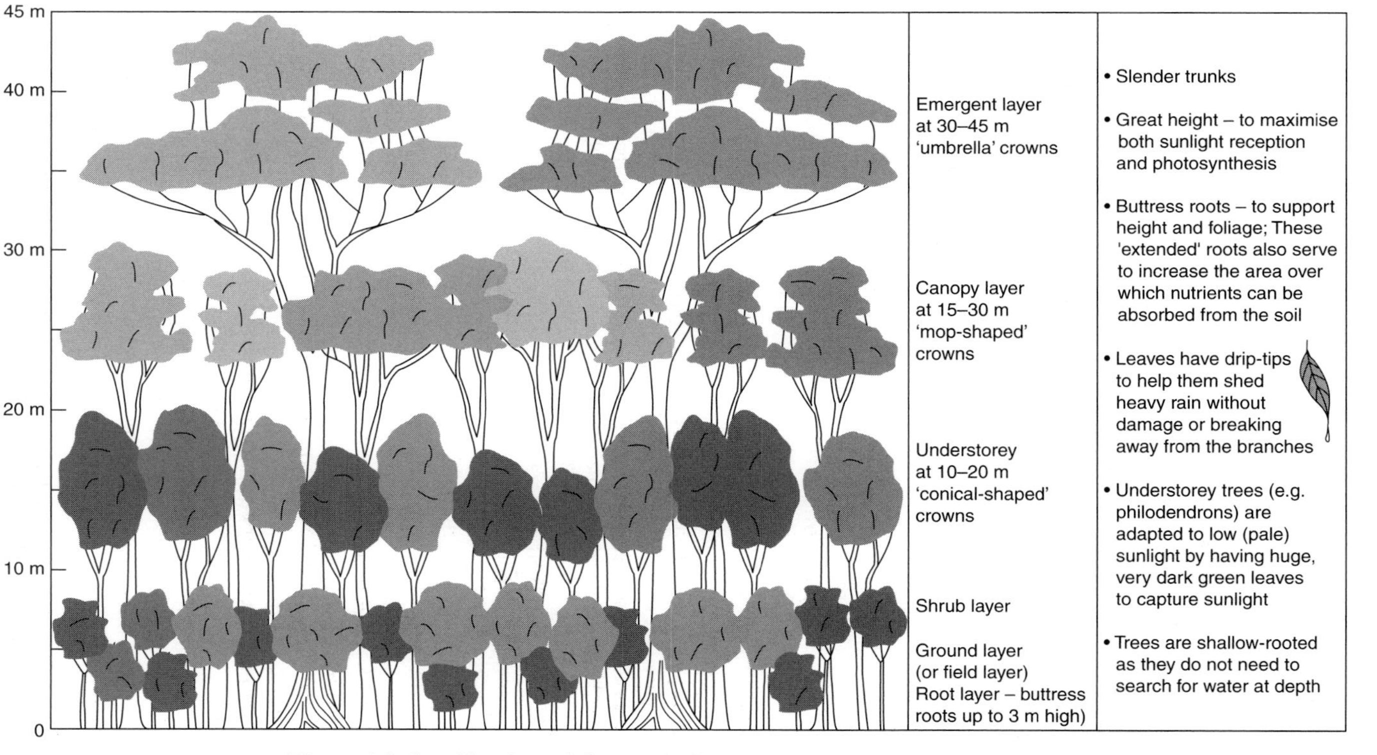

Figure 4.3 Stratification of the tropical rainforest and vegetation systems.

- Epiphytes (e.g. orchids, bromeliads and some ferns) which use the trees as climbing-frames to reach the sunlight; they take moisture and nutrients from the atmosphere, not from their hosts.
- Lianas, which are a type of climbing vine, exist in symbiotic relationship with the forest trees. Most tropical rainforest vegetation is overtall, top-heavy and shallow-rooted, making it easily uprooted. When lianas reach the canopy layer, they continue to grow through the branches of neighbouring trees anchoring their host tree to its neighbours. This gives the trees added support during strong winds, preventing them from being toppled.
- A variety of saprophytic plants at ground level which feed on dead organisms; these are so effective that they are able to break down dead plant and animal matter within 24 hours.
- Pollination is usually by insect, bird or bat. Because of the dense vegetation cover winds are often too light within the forest to pollinate trees, particularly as individual trees of the same species are so far apart. As a result, many trees have evolved very colourful, scented flowers to attract pollinators.
- Trees located along river channels that have adapted to several months' submergence as rivers may rise by as much as 15 m during the wettest season.

Beneath the main forest, there is little vegetative ground cover. The perception that rainforests are dense, impenetrable jungle is largely inaccurate, as it is only along river courses and in forest clearances that sunlight can penetrate to the forest floor and foster growth. Elsewhere, there may be patchy vegetation cover but more generally the forest floor is covered by decaying detritus.

While different vegetation species have evolved within distinct niches so as to reduce competition, individual species remain hugely dependent on one another. For example, the brazil-nut tree only flowers for a month each year. It relies on pollination by insects (usually bees) but depends on other plant species to flower at different times throughout the remaining 11 months of the year in order that the bees can survive to complete the next pollination cycle. After the brazil-nut tree has fruited, its nuts fall to the ground, where the only animal capable of opening the seed case is the agouti (a guinea pig-like rodent). The agouti buries some nuts and its 'forgotten' ones stay in the ground long enough to germinate. Therefore, the nut tree's continuing existence is totally dependent on both bees and agoutis. Such combinations of specialisation and interdependence increase the overall fragility of the biome. Because physical conditions remain so constant, specialisations have evolved to such a degree that even minor environmental disruption cannot be accommodated and individual habitats can easily be destroyed.

d) Soils

The tropical rainforest biome is so large that a single 'archetypal' soil does not cover the entire region; however, it is possible to highlight a range of characteristics that are applicable to most equatorial soil profiles (see Figure 4.4a). The majority of equatorial regions coincide with some of the most geologically (tectonically) stable regions on earth and remained unaffected by the Pleistocene glaciations that had such a profound effect on the development of soils in the temperate latitudes. This means that rainforest soils are ancient, in some places very deep (up to 30 m) and have been subjected to prolonged, intensive weathering. In some rainforest areas, there is evidence of several previous soil profiles in certain areas, indicating that denudation and replacement have occurred throughout geological time. Rapid and intensive weathering across such long periods and under hot and humid climatic conditions means that tropical soils often lack nutrients and are, therefore, relatively infertile.

Although Figure 4.4(a) shows a distinctive tropical soil profile, many tropical soil profiles are much less clearly defined. It is often difficult to distinguish where the bottom of the soil ceases and the upper reaches of the regolith commence. A further difficulty is that most equatorial soils have poorly distinguished horizons due to heavy leaching. As a result of leaching, most tropical soils are ferrallitic. Although the dense rainforest is able to offer the underlying soils some protection from the effects of tropical downpours, it cannot protect the soil totally from the effects of heavy, recurring precipitation.

Under such hot, humid conditions, clays within the soil break down quickly and silicates are rapidly leached down to lower levels; the minerals of iron, aluminium and manganese, however, are relatively insoluble and the sesquioxides of these minerals remain in the upper layers. This process of iron/aluminium enrichment is known as ferrallisation – from the chemical formulae for iron (Fe) and aluminium (Al). Ferrallitic soils are also known as latosols and the accumulation of iron oxides give them their characteristic red (or reddish-brown) colour. In very damp areas, the iron may be hydrated and the soils assume a more yellowy or reddish-yellow appearance. Although distinct horizons are often absent, a characteristic **laterite** horizon often forms when sesquioxides are concentrated by groundflow water. Laterite remains soft while moist, but can become extremely hard after drying out (as can happen after deforestation), so hard that it is used as road surfacing in many LEDCs.

The year-round growing season ensures a heavy and continuous input of leaf litter. This, combined with hot, humid conditions facilitates rapid nutrient recycling which is able to offset some of the effects of leaching by continuously releasing bases into the soil. This reduces the acidity of the soil.

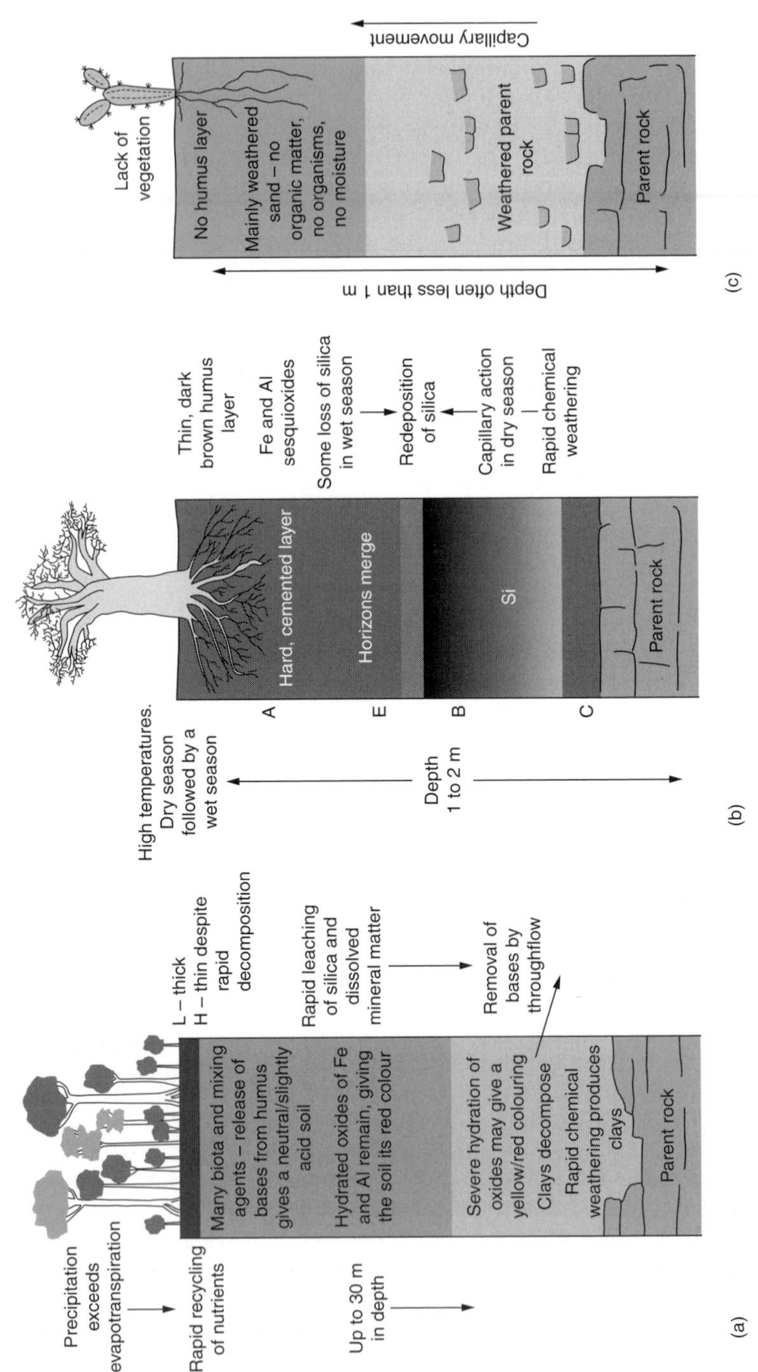

Figure 4.4 Tropical soils profiles: (a) ferrallitic from a tropical rainforest, (b) ferruginous from savanna grassland, (c) grey desert soil.

Ferrallitic soils are infertile. Not only are they ancient soils, which have been subjected to intense chemical weathering across time, they now receive little input of new nutrients. This is the result of a novel adaptation to the nutrient cycle (see paragraph below) whereby nutrients are stored in the biomass (particularly the root system) rather than the soil in order to reduce loss by leaching.

Although the volume of litter falling to the ground is high (on average, over 10 t/ha/yr), its rate of decomposition under humid rainforest conditions is so rapid that the surface litter store is relatively small. As stated earlier, the soils supporting rainforest growth are neither fertile nor nutrient rich. The nutrient cycle in such areas has adapted to offset the effects of heavy leaching and, although the soil contains a wide variety of potential decomposers, the main agents of litter decay are often fungi. These live in a **mycorrhizal association** with the tree roots; a type of symbiotic relationship of benefit to both parties. Nutrients pass directly from the fallen litter into the root network via the decomposing action of these fungi, a process that almost completely avoids the need for soil storage of nutrients. This effectively makes the tropical rainforest nutrient cycle a 'closed system' as so few nutrients are lost by leaching provided that the forest cover remains intact (see Figure 1.9b and c, which show a nutrient cycle before and after deforestation).

Rainforest biomes are the world's most effective converters of the sun's energy. This is largely because the high rate of photosynthesis within the tropics but is also because the trees are able to grow continuously. The table in Figure 1.8 shows comparative productivity data for all the major biomes described in this book.

e) Animal life

It is less easy to generalise about the characteristics of rainforest biome animals, partly because of the diverse array of species, but also because of the many specialisations and niches adopted by animals in order to survive among hundreds, if not thousands, of competitor species. It is estimated that over one-fifth (i.e. more than 600) of the world's bird species, 90% of all primates and over 50 million different species of invertebrates live within the rainforests. Ecological niches can take many forms but some do appear to be common to all rainforests:

- Adaptation to life within a distinctive 'layer' of the forest.
- Adaptation to day-time or night-time activity; this allows more animals to graze and hunt within the same habitat by 'sharing' the time available for feeding.
- Adaptation to feeding on a single food type, e.g. a single variety of nut. Toucans and parrots have developed strong beaks to allow them to open fruits and nuts that have evolved tough shells as protection against predators. Parrots have also adopted ways of coping

with the toxins contained in many nuts; they 'graze' daily on clay river banks, utilising the silica in the clay to neutralise these toxins.
- Most animals live in very limited areas, e.g. the Maues marmoset monkey is endemic to only a few square kilometres of Amazonia.

Most, but by no means all, of the rainforest's animals are arboreal and, of these, the greatest concentration occurs in the canopy layer. Very few animals have adapted to life in the emergent layer other than birds of prey, since at this height it is exceptionally windy, making life hazardous for mammals. Within the canopy, however, there is a rich source of shelter and food in the form of leaves, flowers, fruits, nuts and other animals. Many animals who make the canopy 'home' never descend to the forest floor; others (such as parasol ants) use the tall tree trunks as a 'super-highway' to the forest floor (and shrub layer) giving them access to feeding stations at various heights. Far fewer mammals and birds are permanent floor-dwellers although tapirs (in Brazil), forest elephants (in Africa), big cats such as tigers (in Malaysia) and flightless birds such as the cassowary (in Australia) thrive at ground level. However, the forest floor and the soil layer beneath absolutely teem with insects, ants and termites. Figure 4.5 shows a typical food web for a rainforest ecosystem, together with details of feeding stratifications within the forest.

f) Human activity

Tropical rainforests are among the most threatened ecosystems on earth and deforestation is one of the global community's greatest concerns at the present time. Consider the following points:

- Tropical rainforests used to cover 14% of earth's surface; today they cover less than 6%.
- At the current rate of clearance, tropical rainforests will have completely vanished in less than 40 years.
- Continued deforestation will destroy (or severely threaten) up to 50% of all plant, animal and micro-organism species.
- A single pond in a rainforest can sustain a greater number of fish species than live in all of Europe's rivers.
- A 10-ha plot of rainforest may contain up to 700 known species of tree; this is equal to the entire tree diversity of North America.
- Scientists estimate that at present we are losing up to 50,000 species every year – many of which have not even been identified by 'Western' science.
- Tropical rainforests hold a 'reservoir' of genes that has the potential to increase crop yields and improve disease resistance through cross-fertilisation.
- Concern about loss of biodiversity is not about the disappearance of individual species, it is about the loss of ecosystem functions. Functions which are critical for keeping us alive – and which do so very cheaply.

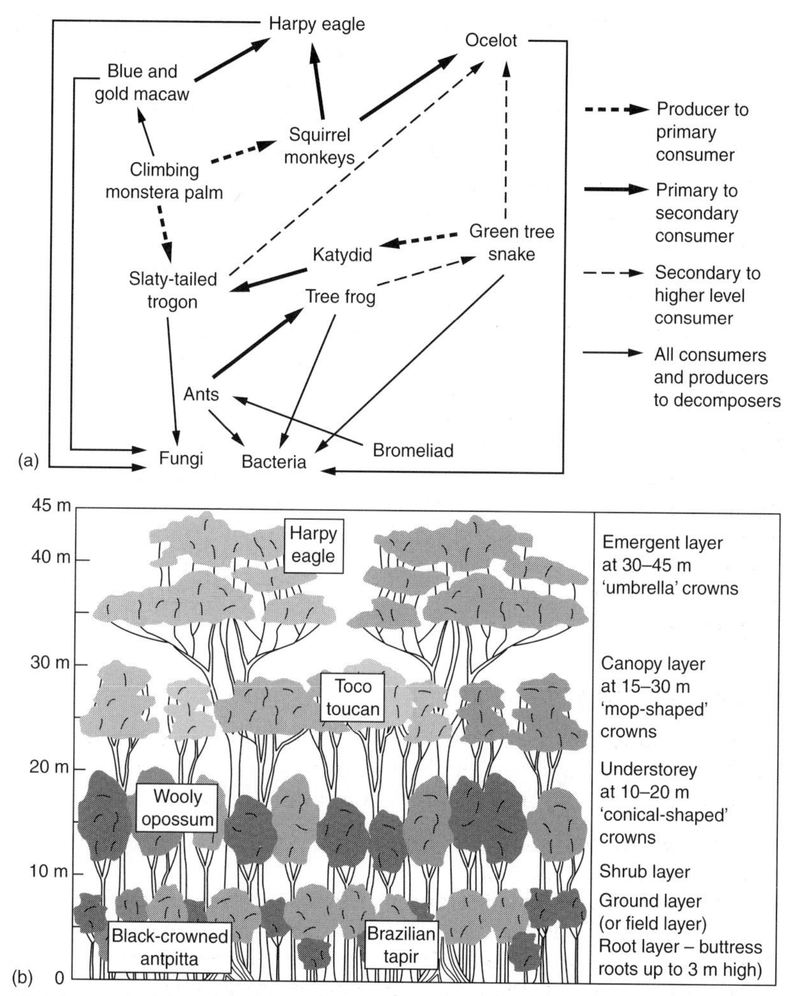

Stratification of rainforest leads to stratified feeding by animal population. Such niches allow different species to avoid (or reduce) competition for resources; this increases the number of species which can co-exist in the forest.

Figure 4.5 Food web for a rainforest ecosystem, together with details of feeding stratifications within the forest.

Deforestation is the result of many factors, but the principal ones include:

- overintensive shifting cultivation, which may be the result of increased population numbers
- logging, which is the cause of the greatest clearances; later the land may be used for ranching or agriculture
- increased collection of fuel wood, which may be the result of

increased population numbers and/or the result of increased charcoal production
- the clearance of land by landless peasant farmers
- clearance for agriculture/ranching/mining
- national or local government incentives to encourage foreign investment
- road, highway and settlement developments.

As geographers, we are concerned about the consequences of both the destruction of the rainforests *per se* and the rate at which destruction is occurring, for a variety of reasons:

- Most academics and scientists agree that deforestation is the result of short-sighted economic activities that value the forest only for its timber.
- Forests generally, and rainforests in particular, are effectively both the source and maintainer of life on earth.
- More that 20% of the world's oxygen is provided by the Amazon rainforest.
- Each year, human activity adds over 6 billion tonnes of carbon dioxide to the atmosphere; about one-third of this is absorbed by the world's forests (mainly tropical rainforests) for photosynthesis. The beneficial effects of this recycling of carbon dioxide have led to forest biomes being referred to as '**carbon sinks**'.
- Increasing carbon dioxide levels are important in both the augmented greenhouse effect and global warming.
- Without tree cover, the hydrological cycle in the tropics is disrupted; both interception and transpiration are reduced and surface run-off increases.
- Currently, one-fifth of the world's fresh water is stored within the Amazon Basin.
- A quarter of all modern pharmaceuticals are derived from tropical rainforest ingredients; over 2100 rainforest plants have now been identified as being active against cancer cells, the rosy periwinkle providing one of the most effective treatments for childhood leukaemia.
- Over 80% of the world's foodstuffs originated in the tropical rainforests. These include fruits such as avocados, pineapples and tomatoes, vegetables like rice, corn, potatoes and yams (which are staple foods for most of the world's population) and other commodities such as latex, black pepper, chocolate, ginger, sugar cane, coffee and many nuts. Cultivation of such crops could help local peoples to increase their income and hence their quality of life.

Most scientists and academics agree that the rainforests are far more valuable if left intact as sources of nuts, fruits, oils and medicines than they can ever be as providers of timber and areas of agricultural activity; current estimates suggest that 1 ha of rainforest is worth approximately:

- £500 if all its trees are felled
- £80 if the land is used for ranching
- £3000 if the land is used for harvesting renewable and sustainable resources.

CASE STUDY: MADAGASCAR

Madagascar is the fourth largest island in the world and a biodiversity hotspot owing to its contrasting natural regions (see Figure 4.6). Because of its isolated island location, 80% of Madagascar's plant species are endemic; likewise, of its 300 reptile species, 95% are endemic and include two-thirds of the world's chameleons. Over 500 of Madagascar's plant species have medicinal uses, the best known being the rosy periwinkle, used to cure Hodgkin's disease, treat leukaemia in children and reduce high blood pressure.

The forest canopy is comparatively low (only 25 m) and includes ebony and other hardwoods and palms together with epiphytes such as orchids and ferns. Some 2175 km^2 of the rainforest lies within the Masoala National Park, created in 1997 and managed jointly by the Wildlife Conservation Society and the Madagascan Park Service to achieve the following targets:

- Using the nearby island of Nosy Mangabe as a safe haven for endangered species such as the aye-aye (Figure 4.7), and the red-ruffled lemur.
- Providing corridors of vegetation to allow animals to migrate between protected areas.
- Educating local people about the long-term economic benefits of protecting Madagascar's valuable natural resources by dissuading farmers from using traditional, but environmentally damaging, slash-and-burn techniques.
- Protecting the coral reefs (which lie within the three maritime reserves) particularly from destructive fishing practices.

About 90% of Madagascar's tropical rainforest has been cleared over the past 1500 years. One reason is the widespread use of slash-and-burn to clear land for farming, a technique that exhausts soil fertility, leaves topsoil vulnerable to erosion and results in severely degraded land. Recently, poor maize, potato and manioc harvests in the south and east of the island have forced local farmers to migrate and scavenge a living within the remnants of the rainforest.

Electricity is generally unavailable, so most people use open fires for cooking and heating; charcoal fuel is more popular than untreated wood because it is lighter to carry, is longer-burning and produces greater heat. Unfortunately, it takes 100 kg of wood to yield only 6–8 kg of charcoal. While logging of virgin

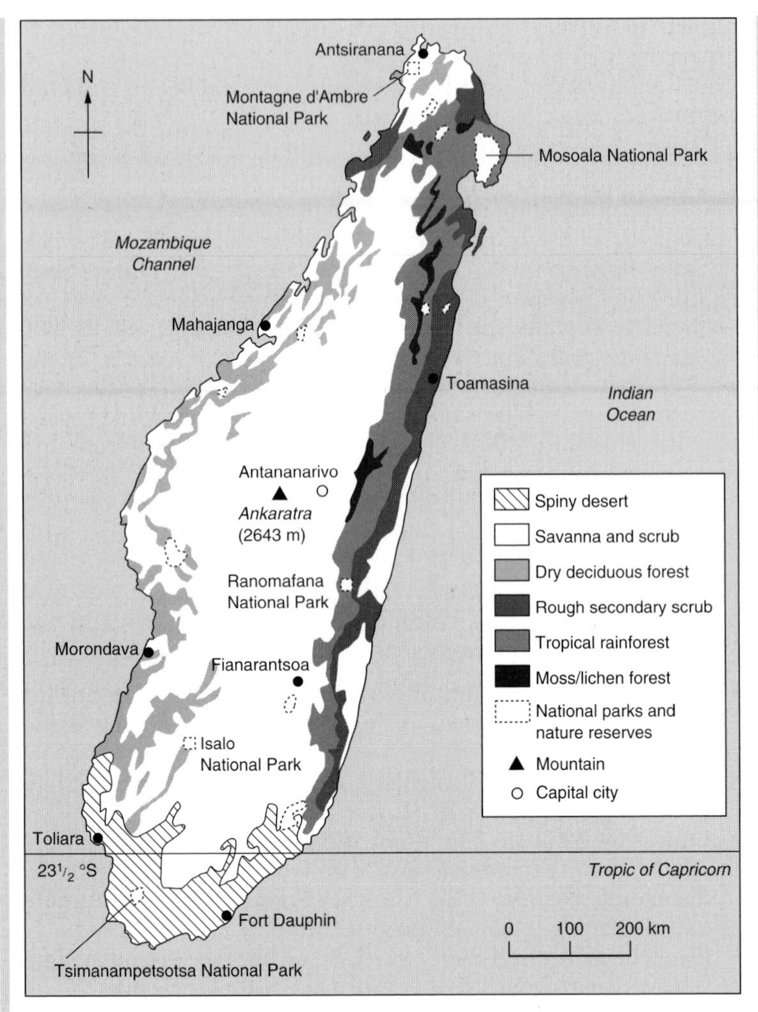

Figure 4.6 A map of Madagascar.

forest is illegal in Madagascar, people in the remotest areas have little fear of being discovered and/or prosecuted and so many trees are still harvested merely for fuel. Some reafforestation is beginning to take place and the Andrew Lees Trust has already planted several thousand saplings and distributed new domestic stoves that are 50% more efficient than the traditional models.

In many ways, Madagascar has been fortunate because deforestation and air pollution caused by forest fires have not occurred on the scale experienced by Indonesia in 1997–8, where over £15 billion of damage resulted from the prolonged droughts caused by the El Niño effect. Fires on such a scale

Figure 4.7 A baby aye-aye, native to Madagascar.

greatly reduce the biodiversity of forests because many species of tree are destroyed but are replaced by only a few, fast-colonising (pioneer) species. However, prolonged deforestation over time has resulted in many indigenous species such as the pygmy hippopotamus and the elephant bird becoming extinct.

In 1986, the Rio Tinto Zinc Company (RTZ) began explorations in the Fort Dauphin region of south-eastern Madagascar for deposits of ilmenite, an ore rich in titanium dioxide which is used as a whitener in the manufacture of paint, paper, plastics, toothpaste and washing powder. Unfortunately, the deposits lie beneath one of the island's few remaining forested areas. This particular area has over 30 plant species unique to Madagascar and is the habitat of the endangered brown-collared lemur. Both the World Bank and the Madagascan government are strongly in favour of the project, mainly because 20% of the shares in the operation would be owned by the government and the income from the investment would help to reduce the country's burden of debt. The decision to start mining was eventually taken on 3 August 2005.

There is great potential for the development of ecotourism in Madagascar. The island has 5000 km of tropical beaches, its climate includes a relatively dry season, it has one of the most diverse ecologies in the world and the capital city, Antananarivo, has large botanic and zoological gardens. The tourist industry is still in its infancy and is therefore well placed to create a truly eco-friendly brand of tourism that could be of net, long-term benefit to the island and its people.

3 The Savanna Biome: A Tropical Grassland

Savanna biomes, like most grassland biomes, are associated with continental interiors. Tropical grasslands are generally located where temperatures are high, precipitation is low to moderate, and drought is a seasonal feature of life. Precipitation scarcity and unreliability means that the effective soil moisture is insufficient to support forest vegetation. However, shortage of rainfall alone does not account for the widespread distribution of savanna grasslands. Towards rainforest boundaries, sufficient rain falls to support tree growth, and other factors such as geomorphology, soils, fire and overgrazing play a key role in the transition from forest to grassland ecosystems. Savanna that is the direct result of climatic conditions and seasonal fire is referred to as 'climatic savanna', while those areas that are the product of soil conditions are known as 'edaphic savannas'. 'Derived savannas' are created through human (or animal) clearance of previously forested areas; while such areas may revert to their former forest ecosystems through recolonisation, this is unlikely because grasses invade and colonise more quickly than trees, so becoming the dominant form of vegetation. One current ecological theory suggests that *all* tropical grasslands represent a plagioclimax succession that is the result of human intervention in former, fragile ecosystems.

a) Location

Savanna grasslands are located in a broad band at about 5–15° north and south of the equator, between the tropical rainforest belt of the equatorial regions and the hot **desert** biomes of the tropics (see Figure 2.2). They cover about 12% of the earth's total land surface, including about 65% of Africa. Other important **formations** occur in Australia, South America and India. The southern Florida Everglades in the USA are an example of a flooded savanna grassland ecosystem.

b) Climate

Savanna grasslands are generally associated with areas beyond the equatorial belt that are affected by seasonal shifts in the ITCZs. Such areas are characterised by:

- high mean temperatures above 20°C throughout the year, with daily highs reaching 36°C, and 11–13 hours of daylight
- generally low-to-moderate total annual precipitation (600–1250 mm), but occasionally in excess of 2000 mm
- a prolonged drought (lasting at least 5 months) that increases the potential for seasonal outbreaks of fire, and creates a cycle of water surplus/deficiency linked to the annual cycle of wet/dry seasons – see Figure 4.1(b).

The drier, cooler season is known as the low-sun period (the season when the sun is not directly overhead) but even then, temperatures remain high and so evapotranspiration demands are great, resulting in an effective soil moisture budget deficit (see Figure 2.3). Rain is often abundant during the wet season; however, the seasonal pattern of rainfall has become far less reliable in recent years. Most rainfall is convectional and is associated with intense electrical storms. These often ignite bush fires, which play a critical role in sustaining the biodiversity of savanna areas. These fires rarely devastate entire communities as most vegetation is **pyrophytic** and birds and animals escape by flying, running or burrowing. However, fires do clear dead and ageing vegetation, leaving a fine coating of ash adding valuable nutrients to the soil and stimulating rapid plant regeneration.

c) Vegetation

Although savanna biomes are generally classified as having grassland vegetation, trees are rarely absent from the landscape; this classification merely indicates the dominant role played by grasses and forbs within the plant communities in such areas. Savanna grasslands are often considered as 'transitional' biomes (or ecotones), marking the gradation from rainforest to hot desert biomes (see Figure 4.8). The change from forest to desert across savanna areas occurs very unevenly. For example, where soil moisture is higher (along river courses or where the water table is high) open grassland may be punctuated by dense clusters of trees, while variations in either soil type or nutrient content may also create localised variations within the general grassland landscape.

Typical parkland ecosystems comprise a mix of wind-pollinated grasses, small broadleaved deciduous plants pollinated by insects, and open-canopy **xerophytic** (drought-resistant) trees. A range of the adaptations to drought, fire and predation which are common in East Africa and Australia is shown in Figure 4.9. In South America and Southeast Asia, however, 'specialist' trees have not evolved. Instead, trees endemic to the neighbouring rainforests have merely adapted to cope with the edaphic or environmental conditions in the adjacent savanna regions. Such communities are often dominated by palm or pine trees with dense undergrowth of grasses. Therefore, savanna vegetation shows distinctive variation from continent to continent, such as:

- The Llanos of Venezuela and Colombia is dominated by rough grassland sustained not by drought, but by the annual flooding of the Orinoco River, after which tree growth is inhibited by long periods of soil saturation.
- The Cerrado of Brazil and Paraguay (the most species rich of all savanna areas) is typified by open woodland, dominated by pine

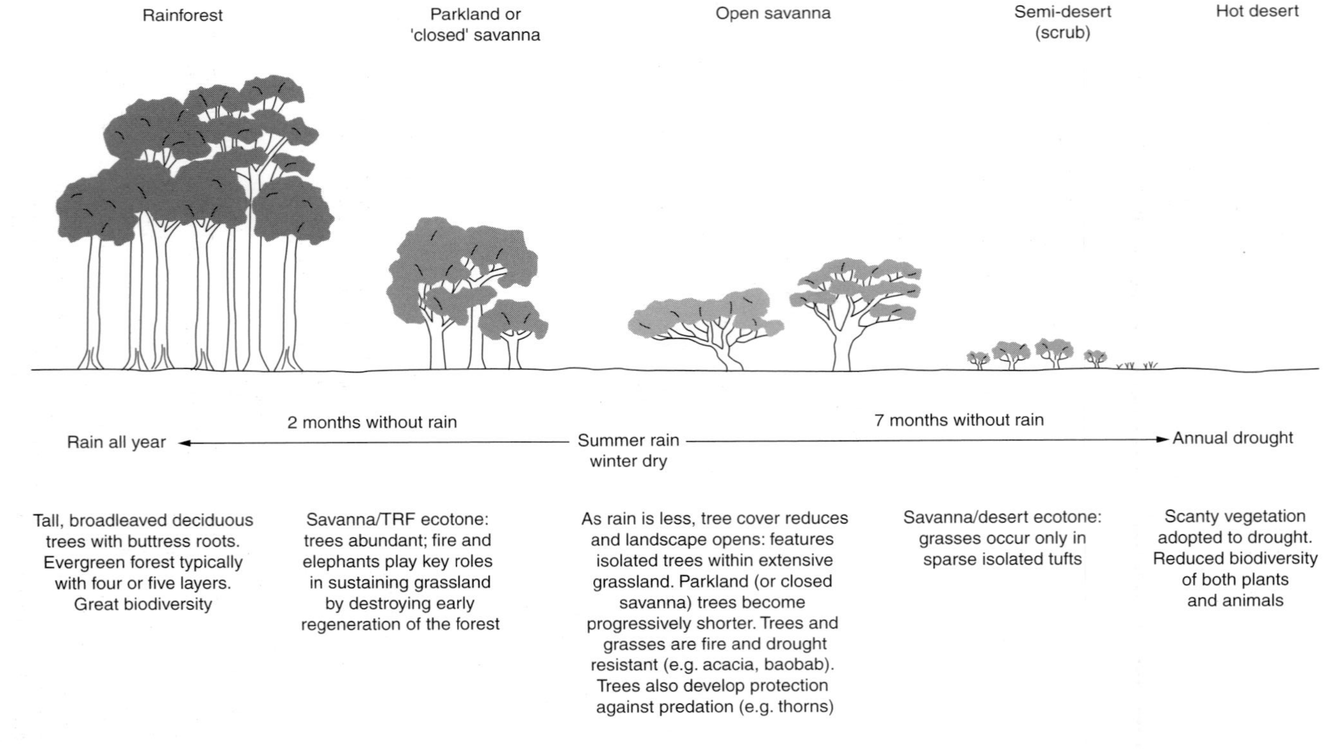

Figure 4.8 Longitudinal section across a typical savanna grassland.

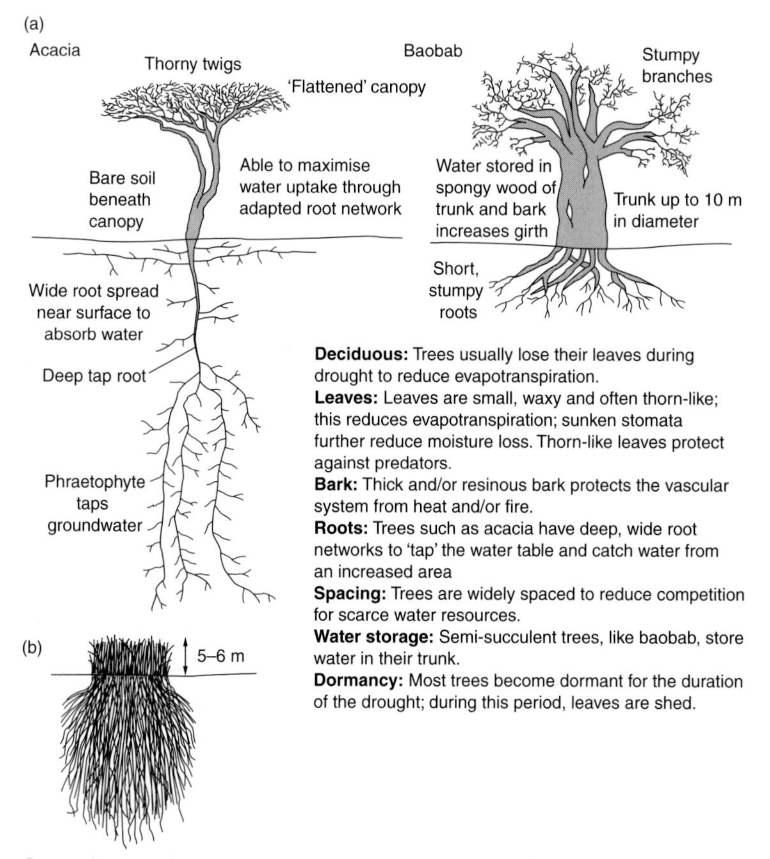

Grasses have a variety of adaptions which allow them to grow successfully under drought/fire conditions:

- They lie dormant until the seasonal rains, then grow vigorously in large tufts or tussocks, often attaining heights of 3–5 m, which means that in some localities they dwarf the more stunted trees around them.
- As the dry season advances, the grasses turn a pale straw colour (from their 'normal' yellow colour) and die back, to remain once again in their dormant state until the next rains.
- most are perennial (live for several years) and their seeds are so drought resistant that they are able to lie dormant for many years. This capability is now increasingly important in many savanna areas where rainfall is becoming less reliable.
- They store food (as starch) and moisture in their root systems.
- They are pyrophytic (fire adapted) as growth occurs from the base of the stem (which is at, or close to, ground level); this means both that when the grasses are damaged in fire, they are able immediately to start to regrow and that the bulk of the biomass is stored below ground. This growth adaption also affords them protection from permanent damage by their predators, provided that they are not overgrazed; as it is being nibbled away, grass is able to continue to grow from its stem base. This is in contrast to trees, which grow from their outer tips; thus, while they may have some adaptations to survive fire, they can be seriously damaged by the actions of browsers who dine on the tips of the branches and leaves.
- Following prolonged drought, grasses have the ability to absorb water much more quickly after first rainfall; they do not need water droplets to reach the finer pore spaces before they can absorb moisture and this therefore means that they are able to transpire fully as soon as rain falls.

Figure 4.9 Vegetation adaptations within a savanna grassland: (a) acacia and baobab trees and (b) grass.

trees with dense undergrowth of grasses; the flora is adapted to cope with unusually high aluminium levels in the soil.
- The pine savannas of Central America are also the product of edaphic rather than climatic variations. Pine trees are one of the only species that can thrive in the drought-like conditions created by sandy, porous soils.
- Eucalypts and spinifex grasses dominate Australia's savanna vegetation but occur alongside numerous species of acacia and one of baobab.

Both naturally occurring and 'managed' fires play a key role in the evolution of savanna ecosystems, clearing and reinvigorating the plant communities. About 50% of the northern savanna regions of Australia are destroyed temporarily by fire every year, partly because eucalypts burn fiercely owing to their high oil content. In contrast, palm savannas in Southeast Asia can survive scorching of the trunk because, being **monocots**, their vascular bundles (their fluid-carrying tissues) are scattered throughout the stem.

Overgrazing also plays a significant role in both developing and sustaining savanna vegetation. This may be due to current farming practices and/or the impact of flourishing populations of herbivores (e.g. African elephants) which open up forest margins to grassland succession by toppling trees along the forest/grassland boundary.

d) Soils

Soils occurring in savanna regions tend to reflect the local seasonal rainfall patterns. During the wet season, precipitation exceeds potential evapotranspiration ($P > pEVT$), there is a net water movement down through the soil and consequently there is a positive soil water budget. Strong downward percolation of water leads to the leaching of soil minerals and clays, with their later redeposition deep within the soil profile. During the dry season, potential evapotranspiration exceeds precipitation ($pEVT > P$), the net water movement within the soil is upward and there is a negative soil water budget. This movement of water, known as **capillary rise**, accounts for the upward movement of silica and iron, both of which are precipitated out of solution just below the soil surface. These processes create leached ferrallitic soils (ferruginous soil), with lateritic strata below the surface. In the long dry season, this laterite layer can become so hard that it restricts drainage and root penetration and even impedes ploughing; it also leaves the soil above it vulnerable to wind erosion.

Like rainforests, many savanna grasslands have developed in geologically ancient shield areas and have been subjected to such extensive weathering that the underlying rocks lack sufficient mineral nutrients to enrich the overlying topsoils.

Relief plays an important role in soil development and subsequent colonisation in these tropical areas. Leaching often removes the clays, humus and sesquioxides from the soils on upper slopes and, by means of throughflow, redeposits these in the soils of the lower slopes, greatly enriching them. This explains why valleys and lower slopes are often densely vegetated by trees whereas upper slopes have only grass and forb cover. As a result of such processes, there are marked catena sequences within savanna biome soils.

The nutrient cycle in savanna areas is regularly interrupted by drought (or by waterlogging in the Llanos of South America). The grasslands provide a high input of potential organic matter, but much of this remains on the surface as leaf litter throughout the dry season. Litter decomposing during the rainy season produces thin, dark layers of humus. Termites compensate for the seasonal cessation of recycling because they feed continuously on the food debris of others and prepare it for final decomposition – a process that takes place rapidly during the wet season. Nutrients which are returned to the soil tend to be stored there until they are utilised, because any stored within the biomass are constantly threatened by the risk of fire (see Figure 1.9d). Leaching is not a widespread problem in clay-based soils, but is a key factor in areas such as Central America where soils are sandy or have sandstone parent materials. However, soil erosion is a major problem in continental interior regions and is the most common cause of nutrient loss from areas within the savanna biome.

The savanna's mean net primary productivity (NPP) (900 $g/m^2/yr$) is considerably lower than that of neighbouring rainforest biomes and is particularly low for tropical locations (see Figure 1.8). This is partly because savanna grasslands support few different species with little stratification and also because decomposition rates in areas subjected to prolonged droughts are significantly lower. Productivity does, however, vary widely from 1500 $g/m^2/yr$ at the rainforest ecotone to less than 200 $g/m^2/yr$ in scrub areas adjacent to hot desert. Although NPP is apparently low, savanna ecosystems recycle nutrients more efficiently than rainforests, in view of the fact that they have less than 10% of the biomass of the rainforests.

e) Animal life

Although savanna grasslands provide a diverse array of habitats, the number of individual species is relatively low (except for insect species); on the other hand, the population numbers for larger animals are often very great. A number of savanna regions are famous for their herbivore species, particularly elephants, giraffes, antelope and zebra in Africa, and kangaroo in Australia. Forming into large herds gives added protection to herbivores in East Africa, especially from their chief predator species such as cheetah, lions, hyenas, eagles and hawks. The larger herbivores in Australia have only one predator: humans. Savanna food chains are very short, often stopping at the secondary consumer level. Herbivores of the savanna (particularly in

East Africa) have evolved highly specialised eating patterns and may be divided into two distinctive 'eating' groups:

- browsers nibble twigs and leaves, e.g. the white rhino
- grazers eat grasses and herbs, e.g. the black rhino.

Herbivores can also be categorised according to the height at which they eat; for example:

- gazelles and wildebeests eat short grasses
- zebras eat longer grasses and plant stems
- elephants eat leaves and the low branches of trees
- giraffes eat leaves and shoots from the tops of trees.

Resources are not only divided vertically (spatially), they are also divided temporally (i.e. at different times of the day or night) with some animals even eating at different times of the year! Such space and time differentiation minimises interspecies competition and is an example of resource partitioning.

During the dry season, competition for water is intense and animals respond to prolonged drought in one of two ways:

- Small animals hibernate, or modify their diet by existing on dormant seeds instead of grazing.
- Birds and most large animals migrate to wetter, more productive areas. Elephants sometimes migrate in this way, but their size and physical strength allows them to tear open the trunks of baobab trees then siphon up the water stored inside them.

Savanna fauna is particularly rich in scavengers such as jackals and decomposers like termites (whose high mounds are an outstanding landscape feature of the East African landscape). Termites play an important role in nutrient recycling and are also responsible for 'grouped-tree grasslands' – where trees grow from isolated, fertile termite mounds.

f) Human activity

Savanna grasslands depend on complex and fragile interrelationships among plants, animals and the physical environment. If any of these components is destroyed or even modified, the whole community is placed at risk. The most widespread human impact on the tropical savannas has been due to overgrazing and trampling by cattle. Some 50 million people in Asia and Africa alone raise livestock in such areas and their populations continue to increase. The result is increased soil erosion, decreased soil fertility and, ultimately, extensive desertification leading to the permanent loss of thousands of hectares of potentially productive land. Traditional pastoralism was nomadic, with herdsmen following the migration routes of wild animals or seeking more succulent grazing for their animals. The twentieth century witnessed a shift from nomadism towards more settled (sedentary) patterns of farming – a change resulting in the overgrazing of pasture close to permanent settlements. However, savanna soils are not well suited to permanent agriculture because they contain insufficient nutrients to sustain agricultural activities. While the use of irrigation and other techniques can make degraded savanna areas productive again, such measures inevitably require financial investment and training facilities beyond the means of most LEDC communities. Indeed, many irrigation schemes in semi-arid areas have accelerated the rate of desertification through salinisation.

Many scientists believe that the introduction of arable farming to savanna areas poses a major threat to world climates. Burning of the grasslands prior to ploughing can release huge quantities of carbon dioxide; doing this has the potential to contribute far greater quantities of that gas to the atmosphere than are derived from clearing and burning the rainforests.

Deforestation is also a problem in many savanna woodlands, particularly in Brazil, where the relatively lush vegetation of the Cerrado is being heavily denuded. This area is home to unique and diverse flora and fauna, including rare bromeliads and orchids. Some damage has resulted from the collection of wood for fuel, but the main threat is from open-cast mining. The Cerrado is also being cleared for agriculture, especially the cultivation of cash-crops such as soya beans and sugar cane.

Perhaps the greatest threat of all to savanna habitats arises from the ever-increasing rates of population growth and urbanisation within tropical regions. For example, Kenya's current population of 30 million represents a four-fold increase since 1957. Most of the resultant urbanisation has taken the form of urban sprawl into the savanna adjacent to towns and cities. Consequently, in some areas, traditional savanna wildlife has become restricted to isolated, scattered pockets of natural vegetation. International tourism poses an additional threat to savanna parklands – especially in East Africa – even though game reserves (e.g. the Masai Mara in Kenya and the Serengeti in Tanzania) have been developed specifically to conserve the indigenous wildlife and provide safe access for visitors.

4 Hot Deserts

Areas are designated as 'deserts' if their total annual precipitation is less than 250 mm; about 20% of earth's surface area falls into this category. All deserts are arid, but they may be sub-divided into hot or cold deserts, according to their temperature characteristics. This section focuses on hot desert biomes, the challenges posed by the differential between water demand and supply and the reasons why these regions often *appear* to be devoid of animal life.

a) Location

Hot deserts are located between latitudes 15–30°N and S, in tropical and sub-tropical areas dominated by atmospheric high pressure for up to 90% of the year. They can occur in continental interiors (e.g. the Sahara, Arabian, Thar and Australian Deserts) or along coasts (e.g. the Atacama and Namib Deserts), as shown in Figure 2.2.

Coastal deserts tend to occur only between latitudes 20° and 30°N and S, along the western edges of continents where the prevailing winds are offshore and hence dry. These coastal deserts are also associated with cold ocean currents such as the Benguela Current off the coast of Namibia and the Peruvian (Humboldt) Current off Peru. The low sea temperatures reduce the capability of the air blowing over them to hold water vapour; therefore such on-shore winds bring little, if any, moisture to the adjacent desert areas.

Inland deserts such as the Sahara are the product of descending air associated with Hadley Cells. Subsiding air becomes compressed and therefore drier and warmer as it sinks, bringing relatively stable, high pressure conditions. **Continentality** (the influence of remoteness from oceans) also plays a significant role in the creation and maintenance of deserts such as the Sahara and those in Australia.

b) Climate

Hot deserts are characterised by:

- Hot summers (mean temperatures of 30–40°C), when the sun is effectively overhead; however, deserts adjacent to cold ocean currents can experience much lower summer temperatures.
- Hot/warm winters (depending on latitude and/or continentality), which may bring rain (see Figure 4.1b).
- A very wide **diurnal range** of temperature. Mean summer day-time temperatures may exceed 38°C but mean night-time temperatures may be as low as –4°C. During the day, high insolation is intensified by the lack of both cloud and vegetation cover exposing desert surface to the sun's full intensity. Conversely, at night, the same absence of cloud cover allows the unrestricted radiation of day-time heat back into space.
- Sparse cloud cover results in clear skies, more or less continuous sunshine and 10–14 hours of daylight.
- The **albedo effect** on surface temperatures. Pale surface colourings such as white salt crusts reflect 40–80% of incoming radiation, whereas darker surfaces (such as outcrops of basalt) may reflect less than 5% of radiation.
- Limited precipitation. All deserts (by definition) receive less than 250 mm of rain per year; however, this generalisation masks variations from 12.5 to 380 mm in individual years. Some areas in the Atacama Desert have received no rain whatsoever in living memory.

- Unreliable rainfall. Rainfall distribution often varies widely from month to month and year to year.
- Rainfall is usually concentrated into short periods of time and takes the form of localised convectional showers.
- High evaporation rates mean that potential evaporation is always greater than precipitation ($pE > P$), creating a negative soil moisture budget; evaporation may be so intense that raindrops are vaporised before reaching the ground. Areas having the highest potential evaporation (pE) are often those with the lowest precipitation (e.g. central Australia's pE is about 30 times greater than its mean annual rainfall).
- Relative humidity rates are very low.

c) Vegetation

While the long hours of daylight, unbroken sunshine and warmth are ideal for plant growth, high surface run-off, low infiltration and high evaporation rates greatly reduce the effectiveness of hot desert rainfall. Further adverse conditions for plant growth include very high day-time summer temperatures, night-time temperatures well below freezing and a lack of protective shade. The concept of 'climax' vegetation is not relevant in many desert areas because rainfall fluctuations ensure that plants are kept in a constant state of change. Many rocky **hamadas** (bare rocky surfaces), **regs** (stony deserts) and shifting dune environments are totally barren; in other areas, the density of vegetation cover is defined by the frequency and amount of precipitation, the depth of the water table and the soil's salinity.

Millions of years of continental separation have led to a high degree of regional plant specialisation (e.g. cacti characterise the North American deserts). A range of xerophytic plant adaptations is displayed in Figure 4.10. Other adaptations include:

- Dormancy during drought. For example, perennial grasses die back to await the next rainy period while annual grasses and plants survive as drought-resistant seeds between precipitation events, having completed their entire life cycle in a matter of weeks. Such annual plants are known as **ephemerals**, e.g. the rose of Jericho.
- Retention of water, by limiting water loss through leaf surfaces. This can be achieved by leaf size, sheen and texture. Small leaves or spines limit the surface area exposed to the sun's heat and also reduce the area available for transpiration. Glossy leaves reflect the sun's radiant heat back into the atmosphere, reducing both their temperature and their transpiration rates. Waxy leaf coatings reduce the amount of moisture that can escape from the leaf's stomata.
- Leaves closing their stomata to reduce moisture loss during the heat of the day and re-opening them for transpiration during the evening and overnight when the air is cool and evaporation rates

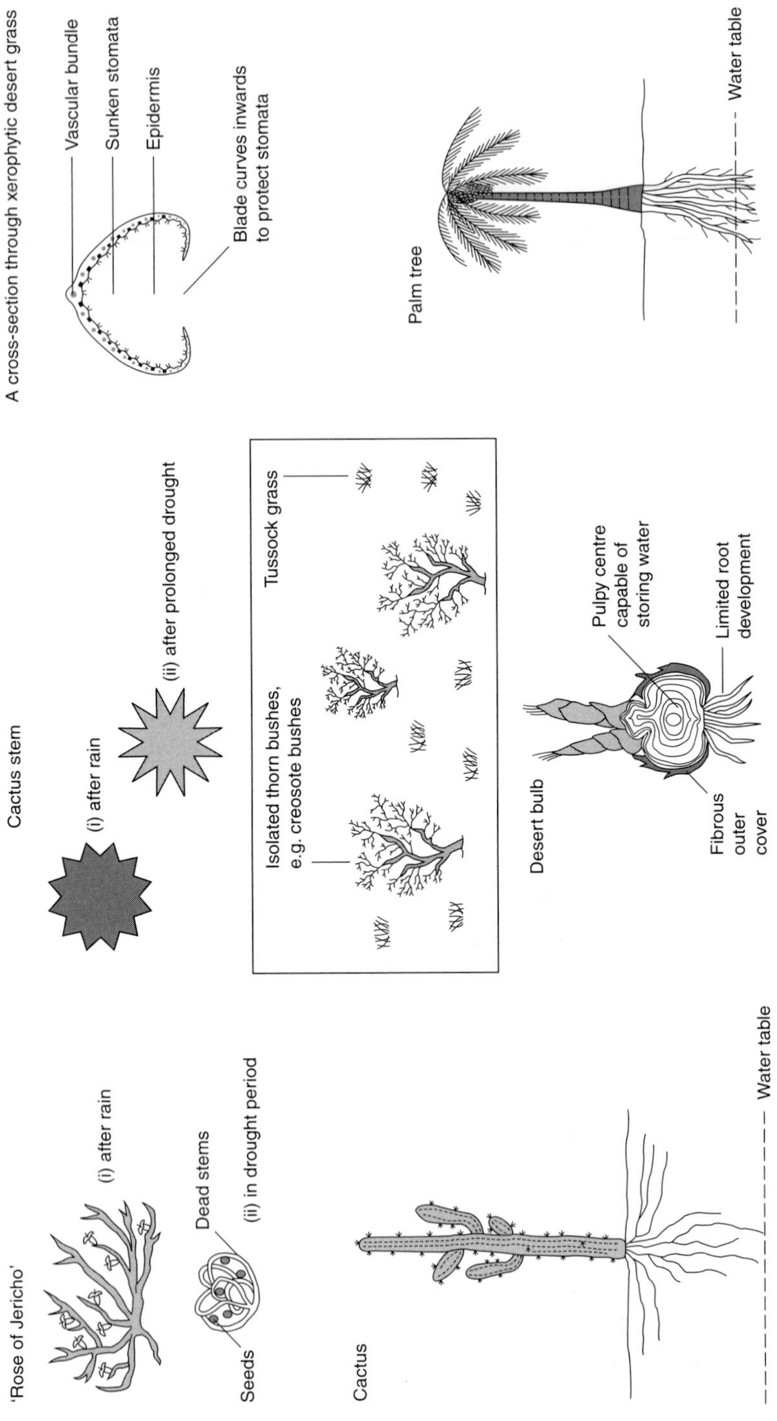

'Rose of Jericho'
(i) after rain
Dead stems
(ii) in drought period
Seeds

Cactus
Water table

Cactus stem
(i) after rain
(ii) after prolonged drought

Isolated thorn bushes, e.g. creosote bushes
Tussock grass

Desert bulb
Pulpy centre capable of storing water
Limited root development
Fibrous outer cover

A cross-section through xerophytic desert grass
Vascular bundle
Sunken stomata
Epidermis
Blade curves inwards to protect stomata

Palm tree
Water table

Hot desert plant adaptations

Other, xerophytic, adaptations include:

- long tap roots which may be up to 50–75 cm in depth; this allows the plant to draw available soil moisture from deep towards the water table. Such plants are known as phreatophytes and are often found in wadi beds where the water table tends to be closer to the surface, e.g. tamarisk and mesquite
- horizontal root system; shallow roots run just below the surface and extend far beyond the plant's canopy in order to collect water very quickly from an extensive area once rains arrive, e.g. cactus
- storage of water in roots, stems, leaves and/or fruit; such plants are known as succulents such as cacti, euphorbia and aloe
- many succulents like cacti have 'corrugated' stems that allow them to expand when water is available and store large quantities; they are then equipped (mainly through spiny leaf structures) to use the stored water very slowly and thereby survive until the next deluge
- ground-hugging form, particularly among shrubs and small trees; stunted bushes are often referred to as 'scrub'
- reduction in leaf numbers per plant to reduce transpiration
- spines are also an effective way to deter animal predators who may attempt to access the succulent's water store
- succulents with spines instead of leaves have chlorophyll in their stems and are therefore able to photosynthesise within the stem
- bushes/small trees may be deciduous (loosing their leaves in the dry season, which is often in summer), e.g. creosote bush
- plants are often lightly coloured as this reflects heat, reducing the need for moisture

Figure 4.10 Hot desert xerophytic plant adaptations.

are low; examples include yucca, prickly pear and brittle bush. The ability to close stomata can also protect against damage during sand storms.
- 'Hairy' leaves provide insulation against both extreme heat and cold; additionally, such hairs trap moisture from the air while reducing moisture loss into dry desert air. In the Atacama Desert, such adaptation allows vegetation to use moisture available in the frequent advection fogs.
- Giving off toxic substances to deter other plants from invading their territory, and hence tapping their water source, e.g. the creosote bush.
- Nutrient storage in leaves allows plants to survive periods of nutrient shortage; such plants are described as being 'replete'.
- Maximisation of water collection; the mulga tree (a type of acacia bush). Unusual among deserts plants, this particular bush has a dense root system clustered at the base of its trunk as well as a large number of upward-growing leaves that capture rain and funnel the water down, along the branches, towards the centre of the tree. This water then drips down to the ground level and is absorbed by the roots concentrated around the bole.
- The ability to survive in saline conditions. Such plants, known as **halophytes**, are not restricted to desert habitats but also grow in locations where salt (which is toxic to most plant species) is abundant, e.g. salt marshes.

Oases occur in desert locations where the water table is at, or very close to, the surface. Sometimes water actually emerges as a spring, while in other places it can be reached by phreatophytic plants with long tap roots. Oasis vegetation is often luxuriant and varied and such areas provide both shelter and water for cultivated crops such as maize, wheat, sugar cane, cotton and date palms.

At the onset of rain, deserts are transformed – almost overnight – into colourful carpets of blossom; even stony and rocky deserts bloom, with tiny plants springing into life in niches between stones and in rock crevices. Grasses and flowering plants predominate, but a number of woody plants including turpentine bush, prickly pear and brittle bush can colonise the most arid of areas.

d) Soils

Soil development within desert areas is retarded by the low precipitation, high rates of evaporation and a lack of organic material; in extreme cases no soil forms at all. Lack of moisture means that the chemical processes necessary for soil development are impeded and that there is negligible vegetation to provide the organic materials from which humus develops. Most desert soils are shallow and coarsely textured because strong winds transport the finer particles away. Because of the aridity, soil biota are restricted to fungi and

actinomycetes (micro-organisms intermediate between bacteria and fungi) whose limited activity is restricted to the infrequent periods of rainfall. Leaching, chemical weathering and the lack of mixing (due to there being few mixing agents such as worms) all ensure that most soils remain relatively infertile. Typical desert soils are zonal aridisols, most being yellow to grey in colour. Lack of moisture restricts hydrolysis and therefore the availability of red-coloured iron oxides to colour the top soil (see Figure 4.4c).

Some desert soils are potentially fertile and can be brought into agricultural use by irrigation – although this can produce problems with salinisation. The most fertile desert soils occur close to oases or alongside perennial rivers such as the Nile and Tigris, where thousands of years of irrigation and cultivation have assisted the soil-forming process and resulted in relatively deep, fertile soils with a high organic and moisture content. Most natural desert soils are, however, affected to some extent by salinisation. The strong upward movement of moisture caused by high rates of evaporation carries salts and bases to the surface in solution. As this moisture evaporates, salts within it are deposited at, or just below the surface, creating salt pans or salt crusts. Soils then become both alkaline (with high concentrations of magnesium, sodium and calcium) and saline. Salinisation can also occur in desert soils due to the underground seepage of seawater (which is then drawn towards the surface through capillary action), excessive water extraction from aquifers and poorly managed irrigation systems. At least 25% of the world's irrigated land has been adversely affected by salinisation, much of it abandoned to revert to desert. However, salt is toxic to most species of plants and therefore such land cannot revert to its former state because only **halophytic** plants are able to colonise saline soils; these will attract exotic species of animal as the new habitat is unable to support indigenous species of wildlife.

Nutrient recycling is almost non-existent and any that does take place proceeds slowly as there is a lack of organic input and very little bacterial activity to facilitate the process. Many plants create their own 'fertile' zone around themselves; here their dropped leaves decay, are recycled and then reabsorbed to satisfy their own nutritional needs. Many plants store surplus nutrients within their stems and leaves, building up nutrient reserves which they can utilise if drought interrupts recycling processes (see Figure 1.9e). The mean NPP for hot deserts is 90 $g/m^2/yr$ with most input coming not from surface vegetation but from the dense root networks underground (see Figure 1.8).

e) Animal life

Hot deserts often appear to be devoid of animal life. Their biodiversity is certainly low, but they are usually inhabited by a variety of small animals, most of which are rarely seen because of their need to shelter from the intense day-time heat. With the notable exception of the camel, there are few large animals because there is insufficient shelter, water or food. Successful animals have usually adapted in both physiological and behavioural ways; physiological adaptations may include:

- Reduced bodily size; most animals are small because food and water are scarce.
- Many animals being cold-blooded; reptiles are ideally adapted to the heat and are desert 'specialists'. Such ectotherms do not need as much food as warm-blooded animals and so can survive without nourishment for much longer. The sun provides heat for them and, if they get too hot, they seek shade in rock niches or under stones. When temperatures fall, reptiles become 'torpid' (inactive) until it becomes hotter.
- Warm-blooded animals adapting to survive with very little water; they get the water they require by eating insects, bulbs and seeds; some animals don't even drink water when it is freely available.
- Salt glands, which allow animals to secrete salt without the loss of water.
- An absence of sweat glands, because sweating involves water loss.
- The concentration of urine and dung, also to reduce water loss.
- Storage of fat in humps or tails, i.e. away from the vital organs; this is because these organs need to be kept cool and fat intensifies the effects of extreme heat.

Behavioural adaptations are much more limited but include:

- Sheltering from the heat of the day; many animals have even become nocturnal. Mammals, in particular, dig burrows 50–60 cm below the surface where temperatures are several degrees cooler. Some species have adapted to living permanently underground, e.g. the naked mole rat, while others, such as the kangaroo rat, shelter during the heat of the day and emerge to feed only at dawn and dusk. Other animals, including birds, reptiles and insects, rest in rock crevices or in the shade of isolated bushes.
- Some animals becoming dormant during extreme heat and/or drought and being active only in the cooler months. Long-term dormancy is called **aestivation**.
- Lizards 'thermal dancing', which involves them standing on one front and the diagonal, rear foot, then changing feet to avoid over-long contact with the hot sand/rock surfaces (see Figure 4.11).
- Some species (e.g. the Arabian oryx) obtaining water from dew, which forms at night, or from advection fog that condenses on their bodies.

Life for all desert animals is precarious because there are few, if any, alternative food sources. For this reason, food chains are short and often as simple as shown in Figure 4.12.

Figure 4.11 A lizard's thermal dance.

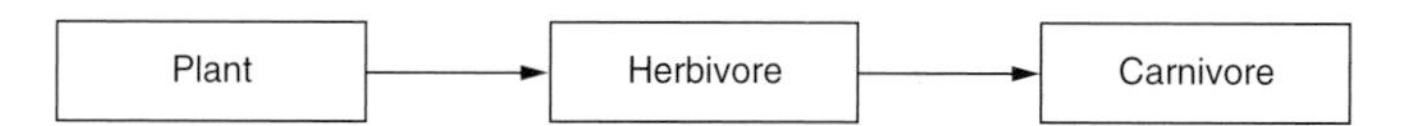

Figure 4.12 A typical desert food chain.

Interlocking food webs are non-existent, owing to the absence of sufficient animal species. Like plants, many animal species are formation (often continent) specific, as shown in Figure 4.13.

Desert	Types of animals
All Australian deserts	Bilby, dingo, kangaroo, marsupial mole
Arabian	Dromedary, civet, fox, oryx, scorpion, vulture
Kalahari	Gazelle, gerbil, hyena, meerkat
Sahara	Desert hedgehog, horned viper, sand fox, mongoose, spiny tailed lizard; gerbil
Sonoran, USA	Barn owl, free-tailed bat, black widow spider, desert iguana, kangaroo rat, rattlesnake, roadrunner, scorpion, turkey vulture
Thar	Dromedary, bustard, jackal, sandgrouse

Figure 4.13 Desert animals and their geographical location.

f) Human activity

Human occupation of hot desert areas has been always tended to be scanty. Some early settlers successfully practised agricultural activities around oases or in river valleys, cultivating cereals, cotton, sugar cane and palm trees; others practised nomadic pastoralism, herding their goats, sheep, cattle or camels between grazing sites. Traditional people such as the Tuareg lived in harmony with the desert, causing little or no damage to its fragile ecosystems. Today, however, where nomadic pastoralists remain this delicate balance has often been disrupted by the introduction of twentieth-century technologies such as diesel pumps to abstract larger amounts of underground water. Not only have such practices depleted aquifer reserves, they have also led to serious overgrazing as a result of increased herd populations.

As traditional lifestyles decline and migration to urban areas increases, prospectors, 'modern' farming methods and 'industrial' activities have arrived to replace them. Because deserts are scantily populated, they are often viewed by more economically developed countries (MEDCs) as ideal locations for toxic (or radioactive) waste disposal and the testing of missiles and nuclear warheads. Yucca Mountain (in Nevada) is the subject of heated debate about the storage of 'spent' nuclear waste. Deserts are also seen as ideal locations for extensive military manoeuvres; however, the heavy vehicles involved easily break the millimetre-thin patina (crust) of ferromanganese, often referred to as 'desert varnish'. This patina is very slow to form but fulfils a vital role in stabilising loose materials on the surface, so protecting them from wind erosion. The 1990 Gulf War culminated in Iraq igniting hundreds of oil wells, an act from which the desert ecosystems will take decades to recover. Other desert usage includes scientific research (many radio telescopes are located in deserts because of their cloudless skies) and film-making, e.g. *Star Wars*. The combination of cloudless skies and abundant sunshine makes hot deserts ideal places to harness solar energy, particularly for LEDCs that cannot afford to import large quantities of fossil fuels. Tourism is also certain to expand, for example, in such locations such as Las Vegas (Nevada) and Mount Uluru (central Australia), both of which are already global-scale tourist destinations. Inevitably, popular sites suffer from land degradation, habitat destruction, the trampling of slow-growing plants and the disturbance of animals' feeding, mating and sleeping patterns. Many smaller, specialist travel companies now promote sustainable ecotourism, particularly in the more fragile, wilderness areas, in preference to destructive leisure activities such as 'wadi-bashing' and off-road rallying. Even so, these remain popular activities, particularly among younger, more affluent males in pursuit of high-risk leisure pursuits.

Many deserts possess valuable mineral resources (see Figure 4.14). Each of these sites creates noise and dust pollution both locally and

Location	Resource
Atacama	Nitrates, copper
Australia	Gold, iron ore, diamonds, silver, copper, petroleum, opals, nickel, titanium
Arabia, Iran, Iraq	Petroleum
Sahara	Iron ore, phosphates, oil
Namib	Diamonds, lithium, beryllium, tin, lead, zinc

Figure 4.14 Desert resources and their locations.

along the transport routes linking the extraction, processing and exporting facilities. During the twentieth century, many desert areas were brought into agricultural production using modern irrigation methods. However, such innovations need to be managed with great care. Excessive irrigation can waterlog soil and 'drown' crops as well as leading to salinisation, which can render the soil agriculturally impotent and lead to desertification (see Chapter 7). Up to 40% of the Sahara, Arabian and Thar Deserts, 25% of irrigated US deserts and 50% of the irrigated land in Iraq was lost in this way during the twentieth century.

The ecology of hot deserts is constantly changing. Wet years allow xerophytic plants to colonise marginal areas, until recurring droughts interrupt their further development. However, recent weather events, particularly along the southern margins of the Sahara (the Sahel), in Australia (which is currently experiencing a drought of epic proportions) and in India, have prompted concerns that the earth is currently experiencing an unusually rapid phase of desert expansion, a process commonly referred to as desertification (see Chapter 7).

Summary

- Within tropical areas, precipitation is the chief determinant of vegetation and, therefore, ecosystem development.
- Rainforests develop where temperature and precipitation are consistently high enough to create both optimum growth conditions and a year-long growing season. Rainforests are characteristically stratified into four or five vegetation layers, which has led to their animal communities adopting similar stratified feeding and living zones.
- Where and when rainfall is intense, tropical soils are subject to leaching and most are ferrallitic as a result; illuviation creates a layer of laterite, which inhibits root penetration and can lead to waterlogging. Most rainforest nutrients are stored within the biomass, so when this nutrient store is removed by deforestation, soil degradation swiftly follows. Rainforest clearances are now a major

cause of global environmental concern. Not only can desertification follow soil degradation, but these forests are one of the earth's major carbon sinks and their loss augments the greenhouse effect.
- Located between the rainforests and the tropical hot deserts are extensive ecotones of savanna grassland. These areas represent a vegetative transition from rainforest to scrub and, ultimately, totally exposed surfaces. Climatically, these areas are a direct response to the annual cycle of alternating wet and dry seasons associated with the passage of the ITCZs. During the long dry season, a negative soil moisture budget develops.
- Fire plays a significant role in the development and maintenance of savanna vegetation and many plant species have adapted by being both xerophytic and pyrophytic.
- Across Africa and in Australia, a number of highly specialist tree species have evolved, whereas, in other savanna areas, the indigenous vegetation has merely adapted to drought. Grasses are the 'ideal' savanna vegetation because they have adapted to withstand fire and drought and survive strong winds, another common feature of these areas.
- As in the rainforest biome, *traditional* human activities across the savanna areas were sustainable and highly sensitive to the fragility of local ecosystems. The adoption of extensive, high-tech agricultural practices has, however, tended to exhaust the soil, leading to land degradation and, ultimately, widespread desertification.
- Hot deserts biomes occur in two latitudinal belts centred on the Tropics of Cancer and Capricorn.
- Precipitation is often unreliable as well as scanty. High evapotranspiration rates therefore result in a constant negative soil moisture budget. One of the most significant features of life in hot deserts is the diurnal temperature range.
- All successful vegetation is xerophytic. In some areas, it also needs to be halophytic because of high salinity levels in the soil. Many species of plant have adapted by developing extensive 'tap-root' systems. Animals are generally small because of the lack of food and many are cold-blooded. Most shelter during the day, becoming active at dawn and dusk – or are totally nocturnal.
- Soil development is retarded by the lack of precipitation and humus. Many desert soils can be developed for agriculture by the application of fertilisers and irrigation water. However, overirrigation often leads to salinisation and desertification. Although deserts are fragile and easily damaged by economic activity, many are underlain by vast mineral wealth, particularly oil. All have the potential to be sources of solar power of global significance.

Student Activities

1. Compare the features of a savanna grassland ecosystem with those of *either* the tropical rainforest *or* a hot desert region.
2. Discuss the extent to which fire is critical to the maintenance and regeneration of savanna ecosystems.
3. Discuss the view that the continuing destruction of the Amazon rainforest is solely the responsibility of the Brazilian people and should not, therefore, be a matter for concern and debate in other countries such as the UK.
4. **a)** Describe *and* explain the concept of 'stratification' in relation to tropical rainforests.
   b) In what *other* ways does vegetation adapt to the challenges of equatorial climates?
5. 'Hot desert areas are virtually devoid of life'. Discuss the validity of this statement.
6. **a)** Explain what you understand by the concept of 'salinisation'.
   b) Why is salinisation an unwanted side-effect of large-scale irrigation schemes?
   c) How can the salinisation process lead to desertification?

5 Temperate and Cold Biomes

KEY WORDS

Active layer upper layer of soil subject to seasonal freezing and thawing in the tundra biome
Boreal (coniferous) forest type of forest occurring in latitudes above 55°N (also called taiga)
Permafrost permanently frozen ground
Podsol soil characteristic of boreal (coniferous) forests; leaching removes minerals and clay from the upper horizons, these are then deposited by illuviation to form hard pan in the lower soil horizons
Prairie grasses tall grasses that flourish in wetter, temperate-interior regions
Reafforestation planting of trees in areas from which previous forest cover has been removed
Solifluction movement of soil down-slope in periglacial areas caused by the summer melting of the active layer
Steppe grasses short grasses that grow in drier, temperate-interior areas
Temperate deciduous forest extensive areas of deciduous woodland occurring in maritime locations within the temperate latitudes
Temperate grassland extensive areas of indigenous grassland occurring across continental interiors within the temperate latitudes – with precipitation of 250–750 mm/yr
Tundra northern continental plains experiencing cool summers and intensely cold winters; permafrost results in poor surface drainage; wetlands are a feature of the summer months

1 Introduction

Across both the temperate and the cold regions of the earth, climatic and edaphic factors interact within a complex web of relationships. While these result in broad, identifiable biomes they also produce a rich tapestry of much smaller ecosystems and habitats, each individually distinctive and unique.

The indigenous vegetation of the temperate regions has suffered as a result of extensive clearance for agriculture and other human activities to the extent that widespread plant and animal extinctions have occurred. Yet more habitats remain under threat today. Widespread clearance is now affecting the northern coniferous forests, while increasing industrial activity despoils vast tracts of fragile **tundra** ecosystem. At a time when the world's focus is primarily on environmental destruction and desertification within tropical areas, it is equally important that the role played by tundra vegetation as one of the world's major carbon sinks is not overlooked.

2 The Boreal Forest Biome

Evergreen coniferous forests may sometimes be referred to as **boreal forest** (meaning 'northern forest') or as **taiga**. These areas represent the most northerly forest biome, extending in a broad band (about 200 km wide) around the globe, poleward from 60°N; there few examples in the Southern Hemisphere because of the absence of large land masses at equivalent latitudes south of the Equator.

a) Location

The coniferous forest biome forms an almost continuous belt across North America (i.e. through Alaska and northern Canada) and Eurasia (particularly in Scandinavia and Russia), covering about 11% of the earth's total surface area (see Figure 2.2). It is restricted to habitats that are climatically sub-Arctic or cold-continental except in cases such as southern Chile, where it occurs as a mountain biome due to greatly increased altitude.

b) Climate

The most distinctive feature of the climate at these latitudes is the harsh winter, during which temperatures fall below –30°C and occasionally reach –60°C. During winter, some areas experience heavy snowfalls, which can lie for months without melting. Snow can, however, insulate the upper layers of soil so effectively that the temperature immediately below the snow-cover is only slightly below freezing, even in mid-winter. In these northern locations, there are minimal moderating effects from the sea and little effective insolation from the sun (with none at all to the north of the Arctic Circle) during the winter months. Winters are dark; immediately south of the Arctic Circle only a few hours of weak sunlight occur throughout December and January. The weather is also characterised by exceptionally strong winds having the potential to damage vegetation and create a high wind-chill effect reducing ambient temperatures still further.

Summers are usually cool. Mean temperatures rarely exceed 15°C although some days reach the low 20s (see Figure 5.1a). A significant plus-factor in summer, however, is unbroken sunshine for up to 20 hours every day; north of the Arctic Circle, summer daylight is perpetual – allowing indigenous plant communities to photosynthesise continuously.

This combination of extreme winters and cool summers leads to large annual temperature ranges. Verkhoyansk (in the Russian

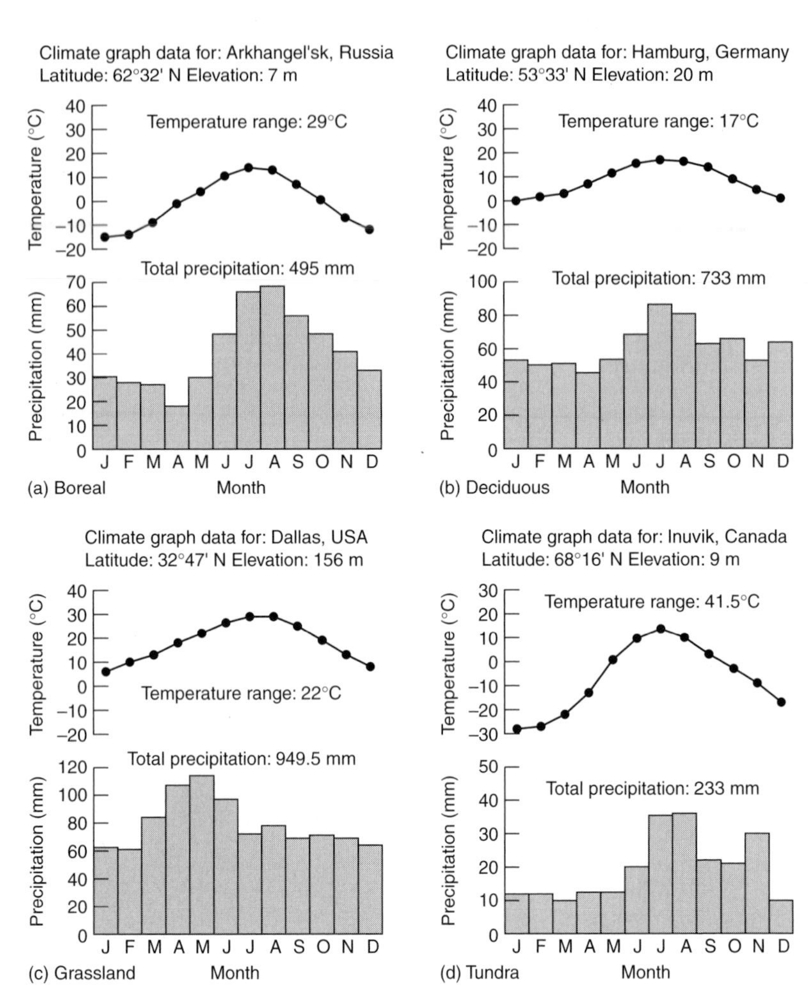

Figure 5.1 Climate graphs for: (a) boreal forest, (b) temperate deciduous forest, (c) temperate grassland and (d) tundra.

Federation), for example, frequently records seasonal temperatures variations as wide as ±36°C. Long, dark winters (with up to 6 months below freezing) and short, mild summers result in an annual total of only 50–100 frost-free days. Such a short growing season is a determining factor for successful colonisation by trees, which require at least 30 continuous days with temperatures in excess of 10°C in order to flourish. Typically, these northern latitudes experience low precipitation totals (again, see Figure 5.1a), with a distinct summer maximum; this is partly because of their distance from the nearest ocean and partly the result of very low temperatures. Although precipitation is low, so too are evapotranspiration rates and therefore the climate is classified as being humid. Most of these areas are not only very cold in winter, but have been glaciated in recent times; in many places the upper layers of soil are still underlain by a thick layer of **permafrost**. This is one reason why snow insulation during the winter months is beneficial for the root systems of plant communities, as it mitigates the effects of the frozen sub-strata.

c) Vegetation

Boreal forest is exceptionally sensitive to local variations in both climate and edaphic conditions and this sensitivity is reflected in the successions and plagioclimax communities we see today. In many forest locations, edaphic factors are the more important inputs in the development of climax vegetation and this is one biome where soil catenas often play a significant role. This is especially true where topographical factors determine the depth of the permafrost layer. More generally, however, vegetation has had to adapt to a combination of low precipitation and a short growing season.

It may be surprising that trees are able to survive such harsh winter conditions at all, let alone dominate the biome community. However, trees are the only form of vegetation that has successfully adapted to survival in such extreme conditions. Unlike the rainforest biome, coniferous forests exhibit only two (very occasionally three) layers or strata. The trees themselves form a relatively dense canopy at about 20 m; beneath this, a restricted shrub layer sometimes exists, populated by dwarf shrubs such as blueberry, cranberry and crowberry. On the ground surface, mosses, lichen and ground-hugging plants such as wood sorrel may be found. Little sunlight penetrates the canopy and there is a thick carpet of undecayed leaf litter on the forest floor; both are detrimental to the colonisation of the lower layers of the forest.

Four main types of coniferous tree occur, each consisting of only a few common species. Spruce, fir and pine are all evergreen conifers, while larch is the exception, being a 'deciduous conifer'. Larch is usually found where the winters are so harsh that even adaptations such as evergreen needle-leaves provide insufficient protection against the tree equivalent of frostbite. In North America, fir and spruce dominate, while in Eurasia the pine dominates, with the Scots pine being the most common species. Although evergreen conifers dominate the vegetation, broadleaved deciduous trees and shrubs such as alder, birch, willow and ash are part of the early stages of both primary and secondary successions. In contrast to tropical rainforests, trees are softwoods and it is rare to find more than two, possibly three, different species per hectare; this is partly because there are so few species adapted to these harsh environments, and partly because the species which successfully colonise localities are often better adapted to localised conditions.

Environmental conditions to which the natural vegetation must adapt include:

- intense cold, often aggravated by intense wind-chill
- extreme variations in temperature
- an effective winter drought caused by the soil being frozen
- a short growing season.

The ways in which evergreen conifers have made the necessary adaptations are illustrated in Figure 5.2. Few mature boreal forests form homogeneous swaths; instead, they tend to be zoned along broadly latitudinal bands which are easily recognisable along any north–south section through the forest; one such transect is represented in Figure 5.3(a).

Edaphic factors, which are often related to the region's periglacial history, cause areas of plagioclimax vegetation. The three most common variations within spruce/fir-dominated ecosystems (i.e. those of North America) are:

- **Bog** (or **muskeg**), which is commonly found in poorly drained hollows and depressions. Ponds of trapped water support thick layers of sphagnum moss on which heathland shrubs and plant species such as cottongrass thrive.
- **Pine forest**, which is common on glacial outwash plains and former dunes where sandy sub-strata significantly reduce effective soil moisture and create drought-like conditions and low-nutrient soils.
- **Deciduous larch forest**, which inhabits intensely cold areas underlain by permafrost. Here the evergreen adaptations to extreme climatic conditions are ineffective. Deciduous birch and aspen may also colonise such areas. Tree cover in these areas is more 'open' than in the coniferous forests and characteristically develops underbush vegetation with shrubs, mosses and lichen.

Current research on the taiga biome suggests that, where spruce and fir are the dominant species at present, they represent only one sere within a 200-year vegetation cycle that is represented in Figure 5.4.

d) Soils

Beneath the trees a deep layer of leaf litter develops that is only ever partially decomposed because of:

- the very low temperatures
- the waxy coatings of the leaves
- the high acidity of the leaves.

The thick carpet of fallen needles creates a physical barrier preventing other plants from colonising the forest floor and so reduces

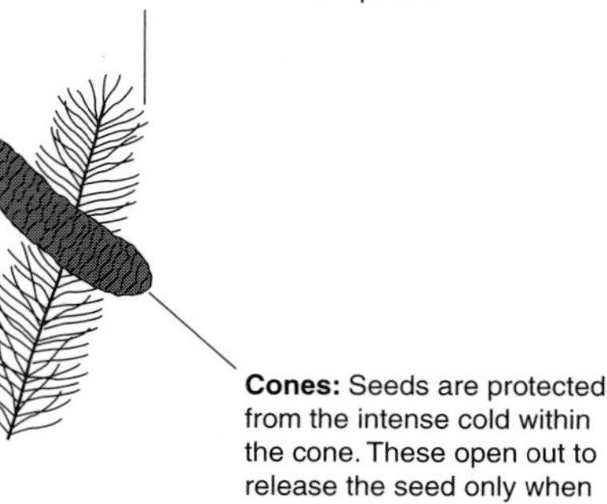

Evergreen: Retention of foliage allows plants to recommence photosynthesis immediately temperatures rise sufficiently in spring. The growing season is too short to grow new leaves from bud

Canopy layer: Tall straight conifers often growing very close together

Colour: Spruce and fir particularly have very dark green foliage which helps to absorb the maximum amount of heat from the sun – allowing photosynthesis to recommence as soon as possible

20 m

Leaf shedding: Leaves are shed only when they are old and no longer functional

Shape: The triangular shape of the trees, together with their short, springy, down-sloping branches, protect the tree from damage by the weight of lying snow in winter. Their shape assists in snow-shedding

Resin congeals if bark is damaged

Field or herb layer: Few grasses and wild flowers grow where light penetrates, e.g. beside clearing; they must be acid tolerant

Ground layer: Thick bed of needles covering the ground; these are very slow to decompose

Wide roots to anchor tree in thin soil

Permafrost (roots cannot penetrate)

Barks: Range in colour from grey to red to brown. They are often ridged or scaled to add greater protection to the inner vesicles of the tree

Roots: Although the trees are often exceptionally tall, they are also extremely shallow-rooted. This means that they can survive above both hard-pan and permafrost

Figure 5.2 Evergreen vegetation adaptations.

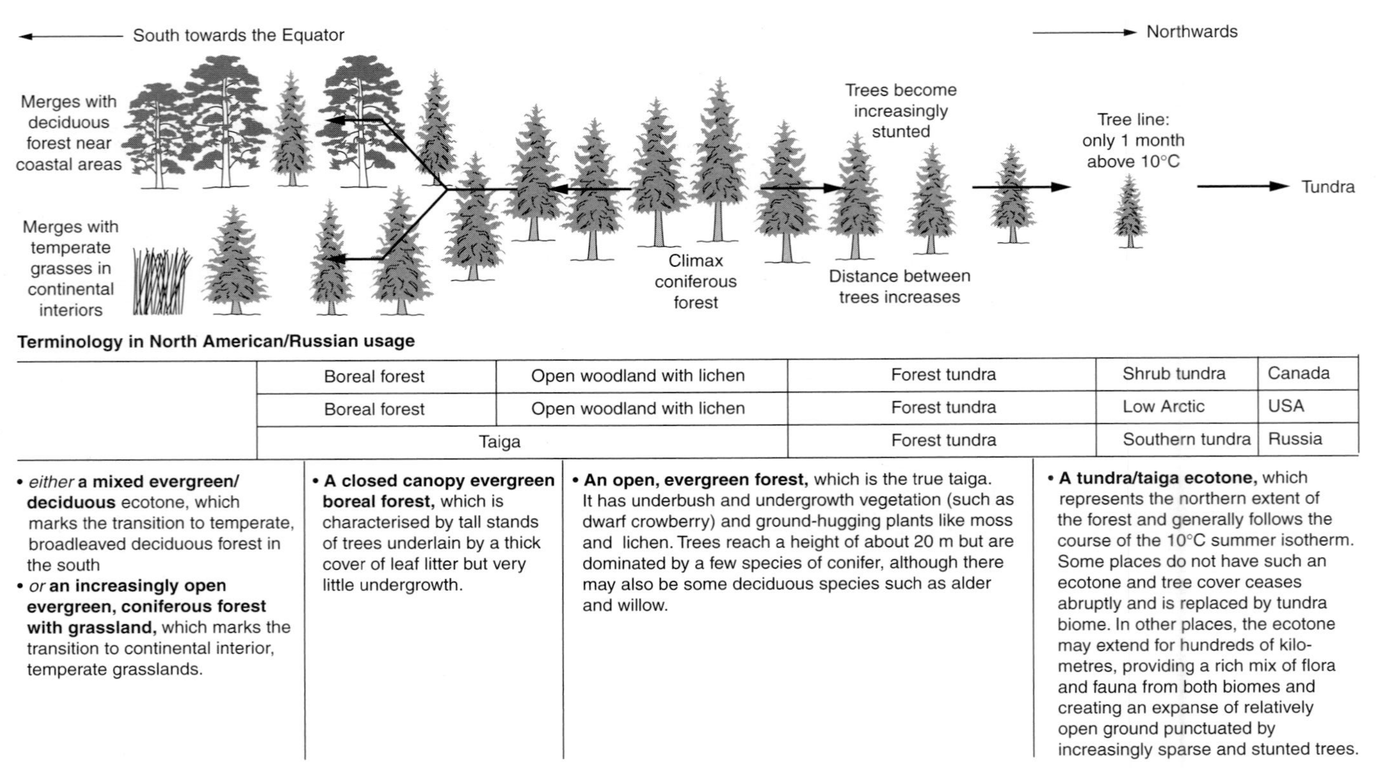

Figure 5.3 (a) Longitudinal section across boreal forest.

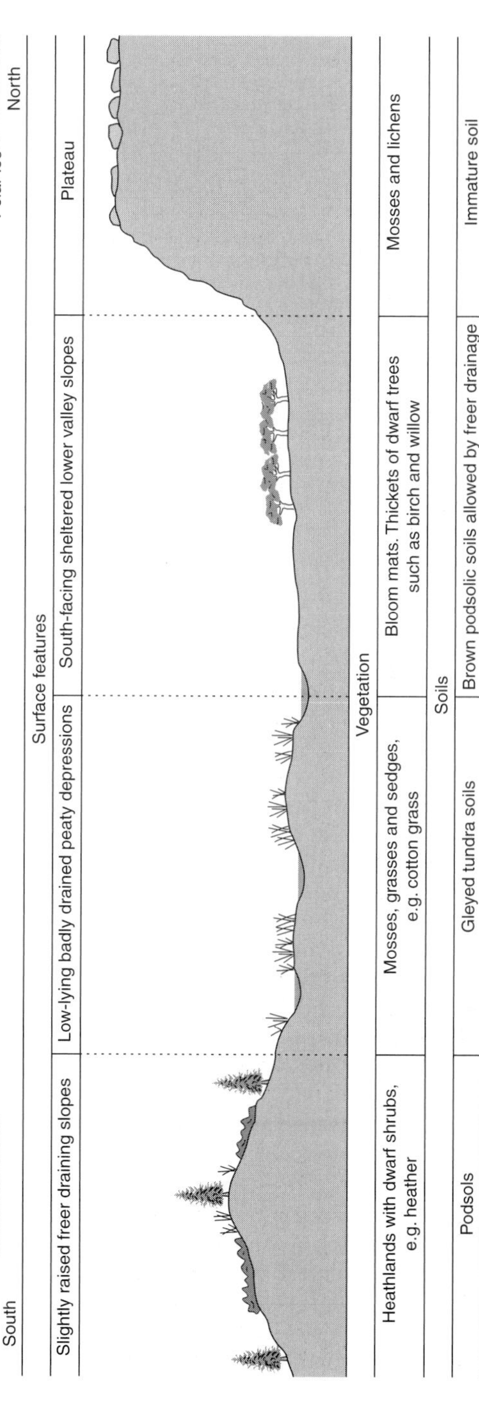

- Middle Arctic tundra is found along flat coastal plains. It is here that rock polygons (patterned ground) are common features of the landscape. Freeze–thaw causes waterlogging of the centre and here sphagnum moss and sedges thrive: around the drier outer ring, forbs and dwarf heathers flourish. Such areas of extensive bog are referred to as muskeg and share similar cold, marshy areas found within the taiga forest.
- Low Arctic tundra makes up the majority of the tundra biome. Here, better drained slopes allow some degree of surface drainage, and generally, the permafrost is significantly deeper. There is a greater frequency of woody shrubs (willow and some birch) and berry-bearing heaths (such as cranberry and blueberry). Here streams frequently flow; along their courses willows and alders may grow to heights of 3 m on south-facing slopes. Further to the south, as the tundra blends into the taiga ecotone, evergreen spruces and firs can be found, although they are significantly less tall than their cousins in the coniferous forest of the taiga.
- High Arctic tundra is the most northerly vegetative community on earth; it is restricted almost entirely to the islands of the Arctic Ocean. Conditions are so extreme that vegetation cover is characterised by lichen and mosses growing on exposed rock surfaces with small pockets of perennial forbs found in protected nooks and crannies.

Figure 5.3 (b) Longitudinal section across tundra.

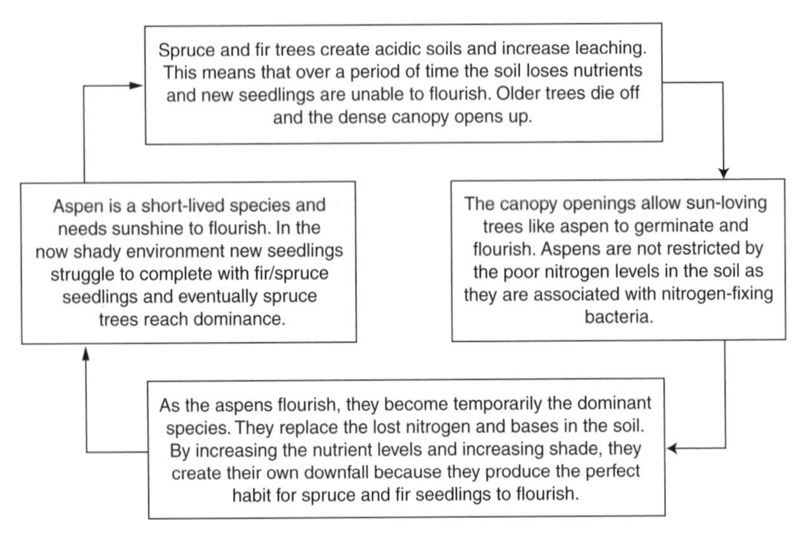

Figure 5.4 The spruce–aspen cycle.

interference competition. Acidic leaf litter decomposes, producing acidic humus that sustains the acidity of the thin, nutrient-deficient soil. These conditions, combined with the very low temperatures and precipitation, give rise to a distinctive zonal soil known as **podsol**. Podsol is a Russian word meaning 'ash-soil'; local farmers, observing the characteristic white horizon below the surface leaf litter, assumed that it was ash residue from earlier forest fires (see Figure 5.5a).

Podsols are most often found in cool/cold climates, beneath coniferous forests where precipitation usually exceeds evapotranspiration; such areas frequently coincide with either glacial outwash plains or regions with acidic parent rock (e.g. sandstones, granite or acidic metamorphic rock). While these latter conditions are not critical for the formation of podsolic soils, they further increase the acidity of the overlying soil. The process by which podsols are formed is called podsolisation (often defined as 'intensive leaching'), but is not confined to boreal forest locations; it also occurs in tropical areas where it gives rise to ferruginous soils.

The needle-like leaves of coniferous trees have thick, waxy coatings and, because few bacteria are active under such cold conditions, they decompose so slowly that accumulation rates far exceed decomposition rates. The humus they provide contains both fulvic and humic acid and is, therefore, highly acidic; such material is known as mor. Additionally, both rain and snow-melt tend to be acidic, further compounding the acidity of the soil. Such high acidity levels facilitate the rapid conversion of iron, aluminium and silica minerals into more soluble forms and the downward percolation of water through the soil profile transports the resulting sesquioxides and clay minerals

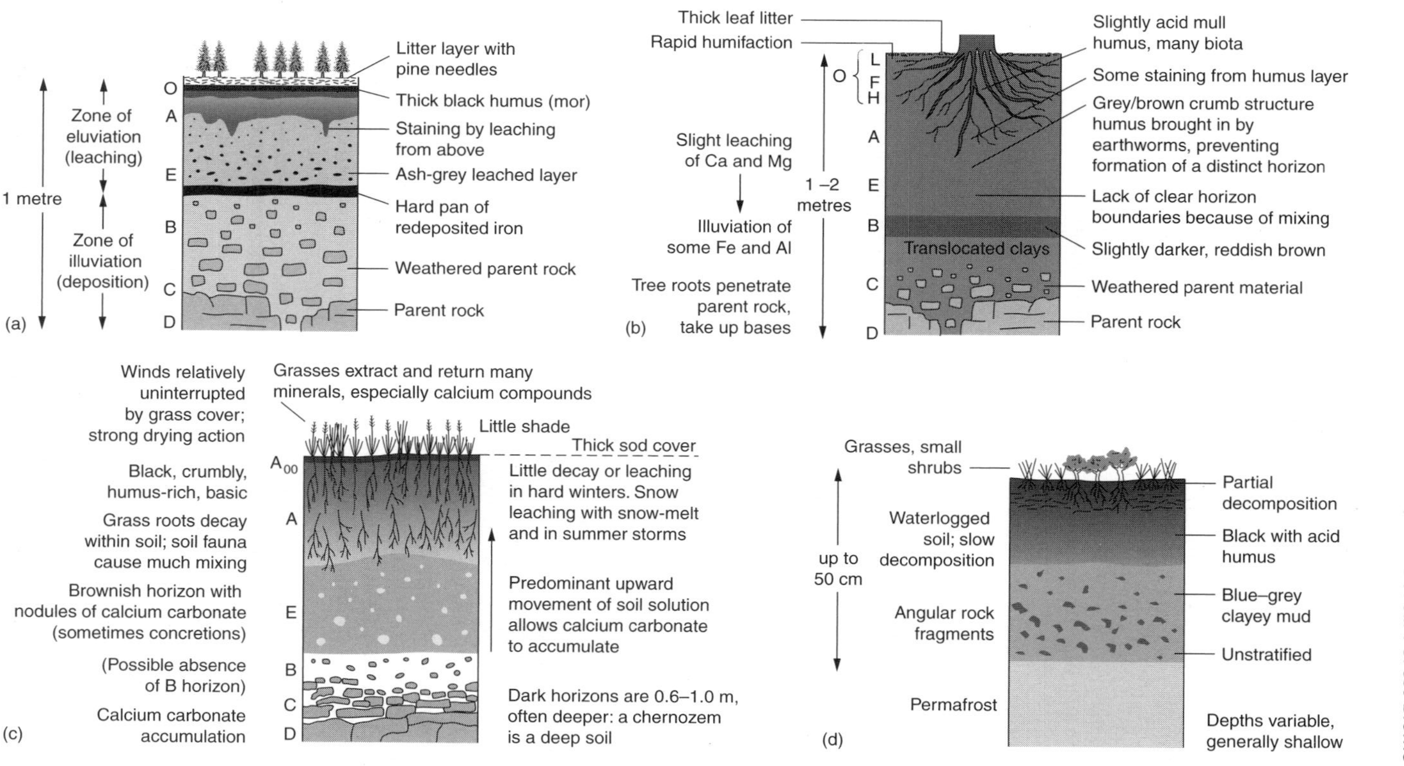

Figure 5.5 Typical soil profiles for (a) podsol, (b) brown earth, (c) chernozem and (d) tundra.

down to the 'B' horizon. Leaching is particularly pronounced during spring, when snow-melt increases water availability. Intensive leaching is referred to as eluviation and produces a characteristic white (or very pale) upper horizon. The cold, acidic soil conditions restrict earthworm populations within podsols so there is little mixing of materials within the soil. This, coupled with the effects of leaching, develops clearly defined horizons within the soil (see Figure 3.7a).

The precipitation of sesquioxides and clay minerals within the 'B' horizon is known as 'illuviation' and creates distinctive dark layering. When organic material is deposited this layering is often black in colour, while the deposition of iron creates a dark red (rusty) coloured layer that forms hard-pan. Hard-pan is impermeable, preventing both the downward movement of percolating water and the penetration of root systems. Ultimately, it can cause long-term waterlogging of the soil above it (creating a **perched water table**) and lead to the development of gleyed (or gleyic) podsol.

Podsols are generally low in nutrient content because leaching removes organic material. Fortunately, conifers require few nutrients (at least in comparison to **temperate deciduous forests**) because of the shorter growing season. The slow breakdown of organic material means that the predominant nutrient store is within the leaf litter layer (see Figure 1.9f). Consequently, if forested areas are cleared, the nutrient store is not automatically removed, as happens in tropical rainforests.

The NPP of the boreal forests is relatively high (see Figure 1.8) considering their low uptake of nutrients and remarkably slow growth rate, averaging about 800 $g/m^2/yr$.

e) Animal life

While most boreal forests have a much greater variety of fauna than flora, the limited plant diversity provides few feeding opportunities for animals; additionally, farmers, ranchers and hunters eliminated large endemic predators long ago to protect livestock and hunting game. Consequently, today's most common animals are seed-eaters (e.g. squirrels and nutcrackers), insects and mammalian herbivores (such as hares) and some elk and moose. Beavers are associated with the aquatic/terrestrial ecotones of such regions. Predators include the weasel family (pine marten, mink and ermine), lynx, bears and wolves. In North America, caribou in-migrate annually from summer grazing in the Arctic tundra in search of winter shelter and food. Generally, birds tend to avoid such dark woodlands, although some seed-eating birds (e.g. sparrows) and omnivores (e.g. ravens) occupy these habitats. In summer, however, thousands of insect-eating birds (especially warblers) migrate to the forests to feed on the hordes of flies, mosquitoes and caterpillars. Most food chains are relatively simple, extending only to third-level consumers. Figure 5.6 illustrates a typical North American feeding web.

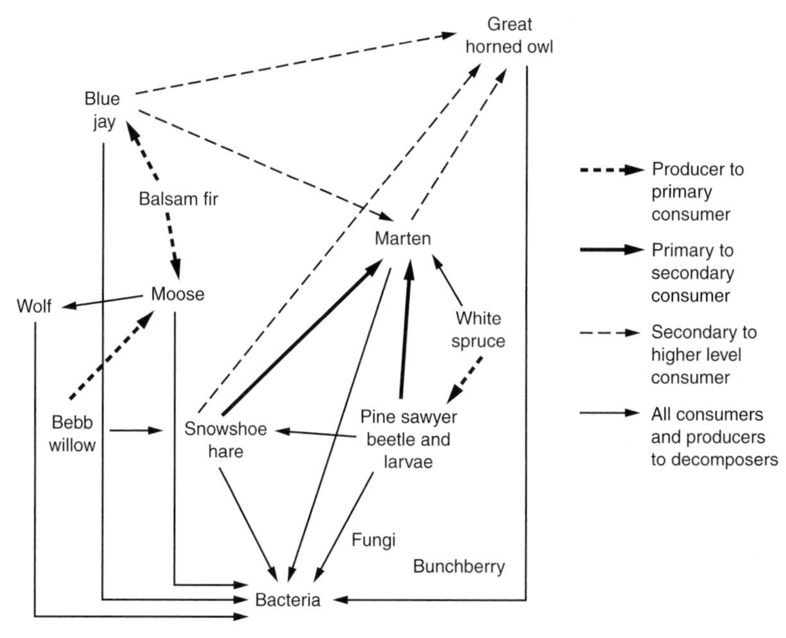

Figure 5.6 Components and interaction in a North American boreal ecosystem.

f) Human activity

Traditionally, the dominant human interest in boreal forests has considered them as sources of timber and animal fur. Timber production involves the deforestation of enormous swaths of forest, to such an extent that many of the remaining woodlands are under serious threat as the world's demand for softwoods continues to increase exponentially. Extensive stands of single-tree species make harvesting the forest exceptionally easy and have accelerated the loss of forest cover to the extent that most European forests have already disappeared. In Canada as well as Europe, **reafforestation** programmes are now an integral part of lumbering operations. Unfortunately, forest regeneration is a long-term process because saplings may take a century or more to reach maturity (in contrast to Britain, where maturity can be achieved within 30 years). Reafforestation usually involves the planting of single species trees in straight, easily harvested rows. This may well guarantee future timber resources, but such woodlands do not replicate indigenous habitats and therefore cannot sustain their original biodiversity. The situation in the Siberian forests is considered in detail in the case study below.

In Europe and Canada, boreal forests have suffered serious damage due to air pollution and acid rain since the late 1950s (see

Figure 7.1), while other forest areas have been lost through submergence as a result of hydroelectric power schemes like the Angara River project in Siberia and the James Bay project in Canada.

While true boreal forests are restricted to the most northerly latitudes, isolated but substantial pockets of locally adapted coniferous forests also occur within mountainous regions in other parts of the world. Boreal adaptations within the Western Cordillera of North America include the giant Douglas firs, sequoia and redwoods. Timber operations threaten these habitats, but the conservation lobby within the USA is more active than its counterparts in Canada and Eurasia and has so far been reasonably successful in preserving these impressive giants, far more so than have the conservationists' attempts to protect the more humble spruces, firs and pines of the true boreal forest biome.

CASE STUDY: EXPLOITATION OF THE RUSSIAN TAIGA

In 2004, three 6-month-old Siberian tiger cubs were tagged with miniature tracking devices; this was just one of a range of measures being taken to sustain this species' population, which had dwindled to less than 400 in the wild. Another north Asian 'big cat' whose low population is causing concern is the Amur leopard. It is the most northerly of the eight leopard sub-species, with only 30 surviving in the wild (although 100 more form part of breeding programmes in European zoos). Conservationists expressed total disbelief in January 2005, when the Russian government announced that it would allow a new oil pipeline to be built through the Amur leopard's last known habitat, the Kedrovya Pad Nature Reserve – a UNESCO biosphere reserve. Evidence suggests that this decision was taken without fulfilling a legal requirement to consider alternative routes and was undoubtedly driven by Russia's need to bolster its ailing economy and to reduce its burden of international debt. Extensive mining poses another threat to the Russian taiga, where reserves of iron ore, diamonds and gold lie beneath the forests. The economic pressure to extract such valuable minerals outweighs any of the conservationists' arguments to protect the forest biome.

Russia has about 760 million hectares of forest, an area 100 times larger than the whole of Scotland. These forested areas are viewed as one of the few ways in which Russia can boost its ailing economy and help to meet the world's ever-increasing demand for softwoods. The Russian logging industry used to be closely regulated, and had effective arrangements for issuing lumbering permits and export licences. Rangers were employed by the Federal Forest Service to ensure that illegal logging was penalised. However, in 2000 this independent service was incorporated into the government's Ministry of Natural Resources, which also happens to be Russia's largest commercial logging company! Theoretically, 47% of the coniferous forests are closed to logging companies, but the pressure on them to meet annual, government-set production targets is intense, especially as they are heavily fined if they underachieve. Consequently, the companies harvest timber from the most accessible locations, whether these are protected or not, because the chances of discovery and prosecution are slim. Currently, Russia invests so little in reafforestation that less than 25% of its felled trees are replaced.

CASE STUDY: THE GIANT REDWOODS OF CALIFORNIA

The redwood (sequoia) trees of north California are the tallest trees on earth, the highest recorded specimen being 112.1 m. Despite their height, they do not have buttress root systems like the emergents in the rainforest; indeed, their roots are less than 3.5 m deep. Instead, the trees usually grow in compact stands, giving each other mutual support by the intertwining of their roots. These giants are exceptionally long-lived; 600–700 years is the norm, but they can survive for as long as 2000 years. Such forests consequently have the greatest biomass of any area on earth.

The coastal strip in California was originally inhabited by indigenous tribes; few Europeans settled there until the Californian Gold Rush of 1848, which prompted a century of continuous logging that resulted in an 85% loss of forest. Tree-preservation campaigns led to the creation of the Redwood National Park in 1968 and enlargement 10 years later. National Park status has provided decades of protection for the groves. However, the Bush Administration hinted recently that these leviathans of the tree world were not so exceptional that they should be excluded from the nation's range of usable natural resources.

3 Temperate Deciduous Forests

Temperate biomes are located within mid-latitude regions which have four distinctive seasons. Across areas where precipitation is higher, forests dominate, but where rainfall is restricted, trees give way to grasslands (see pages 110–115). Temperate deciduous forests are characterised by their seasonal colour change, which marks the onset of the autumnal leaf fall. Most areas that once supported such forests have been cleared for agriculture because they have fertile soils and a long growing season; today very few areas of virgin forests survive.

a) Location

The biome occupies three main formations within the 'mid-latitudes' (40–60°N and S):

- eastern North America, in both Canada and the USA
- Western and Central Europe
- north-eastern Asia, in China and Japan.

Temperate deciduous trees are not entirely absent from the Southern Hemisphere, but only form restricted ecosystems in Australia, New Zealand's South Island and southern Chile (see Figure 2.2).

b) Climate

The mid-latitude regions exhibit wide climatic variations. They are influenced by both warm and cold air masses, resulting in very changeable weather patterns. Continental and eastern seaboard areas experience wide seasonal temperature ranges of ±30°C, while the western margins have warmer winters (with mean temperatures up to 5°C) and cooler summers (averaging 18°C). Mid-summer has long periods of daylight, but winter days are short and the sunlight lacks intensity. Precipitation falls throughout the year and snow is common in winter (see Figure 5.1b). Distance from the sea may increase winter snowfall but reduce the total annual precipitation to 750–1500 mm/yr.

These regions have a long, warm growing season of over 120 days, with at least 5 months frost-free. As rain falls throughout the year, overall precipitation exceeds evapotranspiration, although soil moisture deficits may occur during summer. Extreme weather conditions are rare.

c) Vegetation

As rain falls throughout the year, it is temperature, not moisture, which determines vegetation cover. Nevertheless, trees shed their leaves in winter in response to the fall in temperature and an anticipated need to conserve water.

Undisturbed forest is rare, but displays great diversity and a single hectare can support over ten different species of tree. The majority of tree species belong to the broadleaved deciduous groups (which require 120 frost-free days each year to complete their growth cycle), although there may be occasional stands of evergreen conifer, such as hemlock, spruce and fir. It is common for individual deciduous species such as beech, elm and oak to dominate in some areas. For example, the south and south-east of England is dominated by beech forests. Vegetation cover varies according to soil type; acidic soils give rise to birch, while alkaline soils support box and maple. Willow and alder dominate in wetter areas near to streams and ponds while oak is the most tolerant of all species and can thrive in most habitats. Shrubs and bushes (e.g. azaleas, mountain laurel and blackberry) together with hundreds of 'wild flowers' form understorey layers and ground cover. Forests are usually stratified (in similar ways to rainforests), although normally only two or three layers develop. Figure 5.7 shows a transect through a deciduous forest, including all five possible strata.

Deciduous vegetation has adapted to cold winters in a variety of ways:

- Development of thick bark, to insulate against low temperatures and wind-chill.
- Winter dormancy, to reduce transpiration and hence moisture uptake while soil water is frozen.
- Leaf fall, to prevent 'frostbite' damage and further reduce water loss. It also minimises the overall surface area of the tree and hence the likelihood of branches being broken under the weight of snow.
- Thin, broad leaves, to maximise sunlight capture and photosynthesis.
- Deep, wide-spreading roots, to maximise soil moisture capture.

Leaf fall occurs when trees deprive their foliage of water in response to diminishing light intensity. The withdrawal of water and reduction in sunlight stops chlorophyll production, the leaves change colour and eventually fall. After prolonged summer droughts trees 'close-down' early and assume winter dormancy much sooner. By early summer, trees are able to photosynthesise fully again and are therefore able to flower and bear fruit. After fruiting, they store energy ready for the following spring.

Dead wood plays a major role in deciduous forest habitats, because over half of all woodland fauna depend on it. It provides microhabitats for fungi and lichen and both food and shelter for invertebrates and mosses. Many woodland birds nest within dead trees and some deciduous tree species such as oak and birch exist in a symbiotic relationship with **mycorrhizae**. These fungal filaments live within tree root systems and aid both the decomposition of litter and the absorption of nutrients; fungal fruits manifest as mushrooms above ground or as truffles within the soil.

d) Soils

Soils are rich in both nutrients and minerals. Nutrients are recycled relatively quickly and the plentiful rainfall weathers parent materials ensuring a constant supply of minerals. Leaf fall occurs every autumn

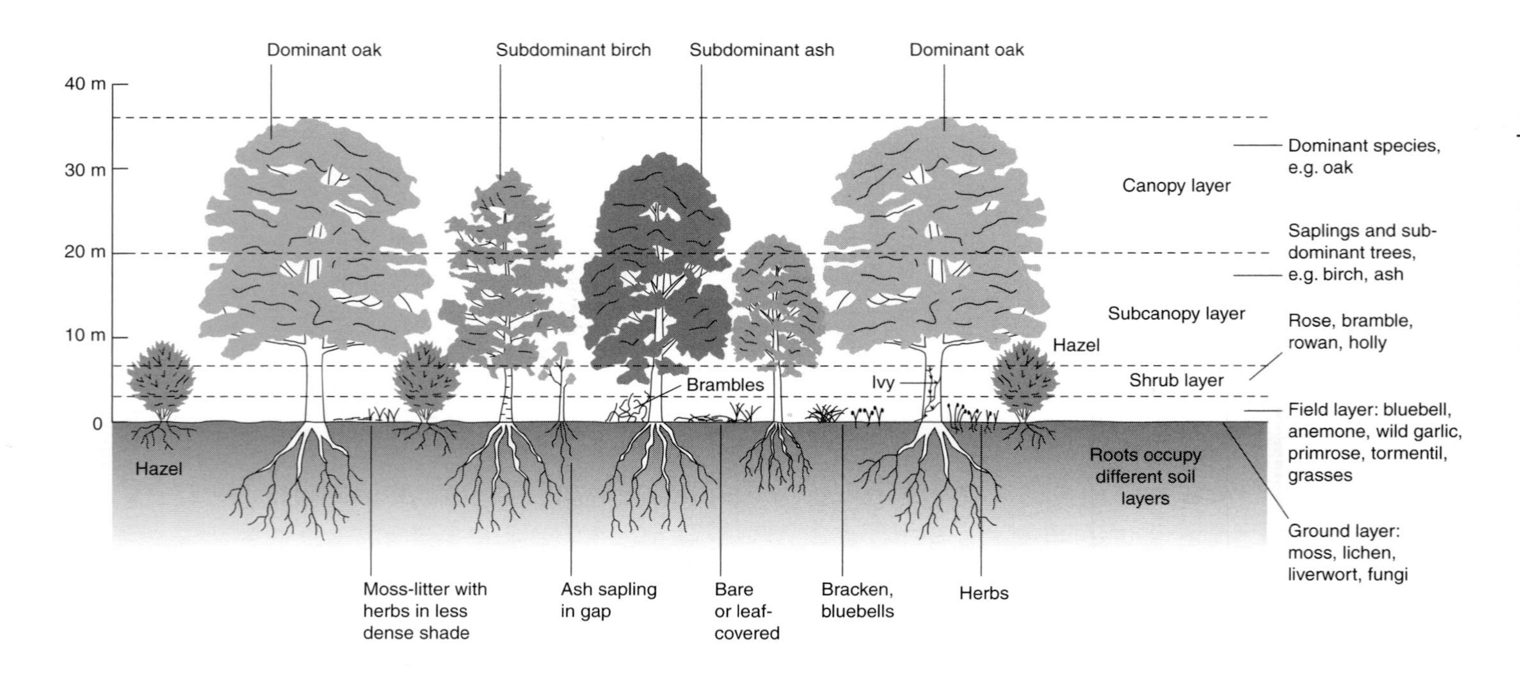
Dominant oak
Subdominant birch
Subdominant ash
Dominant oak
40 m
30 m
20 m
10 m
0
Canopy layer
Subcanopy layer
Shrub layer
Hazel
Brambles
Ivy
Hazel
Roots occupy different soil layers
Dominant species, e.g. oak
Saplings and sub-dominant trees, e.g. birch, ash
Rose, bramble, rowan, holly
Field layer: bluebell, anemone, wild garlic, primrose, tormentil, grasses
Ground layer: moss, lichen, liverwort, fungi
Moss-litter with herbs in less dense shade
Ash sapling in gap
Bare or leaf-covered
Bracken, bluebells
Herbs

Stratification

- At the top is a canopy created by the taller species (up to 40 m) questing for maximum sunlight. The canopy cover is generally complete in the summer, but is rarely dense; more usually, it provides dappled shade to the areas beneath. In spring, uninterrupted sunlight can reach the forest floor because of the absence of foliage. Many of the flowering species (e.g. the horse chestnut) rely on insects for pollination.
- Beneath the canopy may be a layer of shorter trees (i.e. trees which never grow any taller) and saplings.
- The third stratum is the understorey or shrub layer which is comprised of a diverse variety of shrubs such as hawthorn and elder.
- On the forest floor are the forest herbs and the berry-bearing bushes such as blackberry. This area is a carpet of bloom in the spring when plants can enjoy the effects of the sunshine reaching them fully before the canopy is refoliated for summer. Where more light penetrates heliophytes (i.e. sun-loving plants) flourish; in greater shade, the sciophytes (i.e. shade-loving plants) thrive. At this level, most flowering plants are perennial, able to survive the winter by dying back in autumn and regrowing in the following spring.
- A fifth and final layer comprises the lichen and mosses which cover the tree trunks.

Figure 5.7 Stratification of a temperate deciduous forest and its vegetation adaptations.

and produces a deep layer of litter that takes about 9 months to decompose. There is an abundance of micro-organisms to achieve breakdown, but the process slows during the winter months. The humus produced is a slightly acidic mull, well mixed into the 'A' horizon by the actions of soil organisms and giving the top layer a characteristically dark brown appearance. The canopy protects the soil from the direct effects of precipitation; however, some leaching does take place, although its effects are restricted by the high clay content of many soils. It is unusual for podsolisation to affect these brown earth soils, which as a result, tend to be free-draining. Soil horizons are usually indistinct, because of the vigorous activity of mixing agents which range from earthworms to mammals as large as moles. Leached clays and sesquioxides are redeposited within the 'B' horizon, which may assume a reddish hue where aluminium and iron accumulate (see Figure 5.5b). Most **brown earths** are relatively deep and able to help to stabilise the tall vegetation cover above.

The NPP depends on the plants' ability to re-establish the process of photosynthesis in early spring. Overall figures for productivity are high, due mainly to the long summer days and high temperatures, and the mean figure is 1200 g/m^2/yr (see Figure 1.8).

Annual leaf fall provides plenty of material for decomposers, soil bacteria, worms, grubs and fungi. As the leaves decay on the soil surface, rainfall may help to wash nutrients down into the soil without waiting for the humus to be incorporated into the upper horizons. Mycorrhizae play an active role in both decomposition and nutrient uptake. Nutrients are stored within the leaf litter itself, in the soil and within the vegetation, although the major stores are the soil, because of tree dormancy in winter, and the biomass (see Figure 1.9g).

e) Animal life

The relatively dense, diverse flora provides a variety of habitats and plentiful food for a diverse range of animals. As in the rainforests, layering of the forest leads to stratification within the animal kingdom. Huge numbers of birds, mammals and insects live within the forest habitats, as well as many amphibians and lizards in damper niches and undergrowth. Snow-melt and plentiful rain create numerous streams and freshwater ponds which also provide ideal habitats for fish, frogs, newts and insects. Insects play a critical role in maintaining these ecosystems because they are the major agents of pollination.

Many large browsers and top carnivores have been hunted to extinction by farmers who perceived them as threats to their crops, grazing land and animals.

Today, smaller mammals such as rabbits, hedgehogs and badgers live on the forest floor, eating seeds, foliage and insects. Other mammals such as squirrels, raccoons and chipmunks have adapted to life in the trees, thereby increasing the range and populations of inhabitants the forest can support. Thrushes, chats, warblers and many other birds thrive as a result of the vast numbers of insects and seeds while predators such as owls and sparrow hawks are also present in large numbers. There are also smaller populations of other native predators including cougars, bears, wolves and foxes.

Animals need adaptations to survive the cold winters. Many of the larger mammals have double coats, with a soft, short layer that traps warm air close to the body and a longer, outer layer giving protection from wind, rain and snow. Smaller mammals often hibernate throughout the colder months, having previously stored reserves of body fat. Other animals (e.g. squirrels) hoard nuts and seeds to last them through the winter. Most bird species migrate and overwinter in warmer climes because their food sources are unavailable during the winter months.

f) Human activity

Globally, very little virgin temperate forest remains. The finest surviving examples of ancient forest are in Poland and the Appalachian mountains of North America. Britain's 'ancient' woodland is barely recognisable from the oak forests that covered vast areas after the last Ice Age. Many of our woodlands have since been 'coppiced', to provide wood for charcoal burning; this process stimulates undergrowth and produces far more low-level habitats than occur naturally. Most upland British forests are the result of recent reafforestation programmes and are composed almost entirely of conifers because they mature in only 30 years compared to 200 years for the native deciduous species. Meeting the needs of a rapidly growing population has been a major factor in the destruction of our temperate forest habitats, particularly for agricultural, industrial and residential purposes.

CASE STUDY: THE DEVELOPMENT AND MANAGEMENT OF BRITISH WOODLAND

For hundreds of years, Britain has been one of the least wooded countries in Europe, yet pollen records show us that, at one time, great deciduous woodland blanketed the whole country from coast to coast. While pockets of **ancient woodland** remain, this is often by accident rather than design. Such ancient woodland should not be confused with **wildwood** (primary wood), the term for the pre-historic forest that developed as a result of primary succession at the end of the last glacial period.

At this time, the earliest tree colonisers were birch, aspen and sallow (all tundra species), which, about 10,000 years ago began to successfully replace the grassland cover. About 8500 BCE, pine and hazel replaced birch, being followed successively by oak and alder, lime and elm and, finally, by holly, ash, beech, hornbeam and maple. Native beech and lime remained restricted to lowland Britain (i.e. as far north as the southern Pennines). By about 7300 BCE most native species were well established and a stable period followed until about 4000 BCE, when anthropogenic influences began to affect the wildwoods. Across Britain, five wildwood provinces have been identified:

- birchwood, in the extreme north and west of Scotland
- pine, across the Scottish Highlands
- oak/hazel, across northern and south-western England and much of Wales
- hazel/elm, in south-west Wales
- lime, found only in England, south of the Pennines (but excluding the south-west).

Across England, in particular, these woodlands formed a tapestry of tree communities with varying combinations of species in response to variations in soil, relief and aspect. Understorey vegetation is more difficult to determine as under-shrubs and herbs produce far less pollen, but bramble, bluebell and wood anemone would have occupied well-shaded areas, with buttercup, 'ragged robin' and 'devil's bit' flourishing where cover was discontinuous.

Today, there are between 50 and 70 native British species of tree, all of which arrived through natural succession. The ancient woodlands also include some naturalised species (imported from overseas) such as sweet chestnut and sycamore, together with a variety of exotic species, e.g. the Sitka spruce and the Norway spruce, which, while able to survive, cannot reproduce. Some species like the Scots pine and the beech are native only to specific localities within Britain (i.e. the Scottish Highlands and south-eastern England, respectively).

There is no evidence to suggest how these primary forests were cleared, although we can deduce that they were not burned, as only one native species (the pine) can be cleared in this way. It is probable that most were felled and that new growth from the root-stump system was then grazed by livestock and wild animals until the tree finally died. The pollen record shows that clearance began about 4000 BCE but occurred only slowly until the Iron Age, when the introduction of the heavy plough enabled farmers to move onto areas of heavier clay soil and the iron axe-head facilitated tree felling. By 500 BCE, over half of Britain's wildwood had been cleared and woodland management techniques were employed extensively to husband the remaining woodland to supply both timber and wood for fuel. This process of woodland management (outlined below) effectively converted the remaining wildwood into what we recognise today as ancient woodlands because it was enclosed, 'owned' and managed in such a way that both the trees and their underlying strata were altered permanently from the native, primary successions.

Woodland management in Britain depended heavily on the natural regeneration properties of native trees; although these can reproduce by seeding, they rarely do so, because most species are shade intolerant, which means that their saplings cannot flourish beneath dense canopies. Beech is particularly shade resistant, but is able to grow beneath an oak woodland canopy and then often takes over. Oak saplings can grow under birch, but birch itself will only grow in cleared areas. The actual process of tree regeneration is still not fully understood, but it is known that most native species are able to regrow from the stump/root system after a tree has toppled. It is this property that is utilised by woodsmen to harvest and regenerate our woodland species.

Pollarding involves cutting a tree 2–4 m from the ground and leaving a permanent trunk, its bole, which then sprouts anew, conveniently above animals' browsing height. **Coppicing** involves cutting the trunk at ground level from which the new growth (its **spring**) can sprout by as much as 5 cm per day. The disadvantage of coppicing is that the new spring is attractive to cattle and sheep and this is the reason why many ancient woodlands were enclosed for protection. Both management practices harvest the new tree growth in a 5–8-year cycle, providing a sustainable and long-term source of wood and timber.

By 1086, according to the Domesday Book, only a small area of wildwood remained within the Forest of Dean, all other woodland being owned and managed either by individuals or communities. Folklore suggests that, from that time onwards, woodlands have been systematically felled, with the cleared areas used primarily for agriculture. However, evidence suggests that, although the area under woodland had decreased, woods were actually being managed intensively and conservatively. Every person and business was heavily wood dependent at that time and so it was important that such a vital resource was well managed.

From the thirteenth century, woodland plantations were introduced (although it was not until the seventeenth century that they became a common feature and even then only in Scotland, because Scottish law made it compulsory) and the technique of pollarding was replaced by widespread coppicing. The eighteenth century witnessed extensive woodland clearance particularly by wealthy land-owners who replaced ancient woodland by open, landscaped 'parkland'. During the nineteenth century, the practice of coppicing virtually died out – mainly because the Industrial Revolution led to timber and wood being replaced as the major sources of fuel and building materials.

The First World War prompted a major change in national woodland management; Britain's sources of imported wood pulp were lost during the conflict and this led to the creation of the Forestry Commission in 1919. Most new conifer plantations were in areas of moorland and heathland, but some ancient woodlands were also subjected to afforestation programmes. While felling increased up to 1945, it was during the immediate post-war period that most ancient woodland was lost. Over one-third of the remaining woods were cleared in less than 25 years owing to deforestation (to supply industry), agriculture (because of the drive to become self-sufficient in foodstuffs) and urbanisation. Had this trend continued, there would be virtually no native woodland left in Britain today. However, very little woodland has been 'lost' since 1975, mainly because threats to ancient woodland are now immediately (and usually successfully) challenged by conservationist groups. Even the Forestry Commission has modified its planting policy and now utilises more native deciduous species so as to regenerate natural habitats. Additionally, tax incentives are now available to farmers to replant and conserve woodland.

4 The Temperate Grasslands

The **temperate grasslands** are generally associated with mid-latitude continental interiors, far from the coast and rain-bearing winds. In the Northern Hemisphere, temperate grasslands cover vast tracts of North America and Eurasia, but in the Southern Hemisphere, they are more restricted.

Almost all such grasslands occupy flat plains or undulating (gently rolling) terrain. Undisturbed, the habitats were home to over 80 species of mammal, 300 types of bird and hundreds of different flowering plants. However, such regions make ideal agricultural land and few undisturbed grasslands remain today. Consequently, some species have already been lost and many others are under increasing threat of extinction.

a) Location

The most extensive grasslands occur between latitudes 40 and 60°N, with considerably smaller communities between 30 and 40°S. The major areas occupy the continental interiors of North America (i.e. the Great Plains of the USA and Canada) and Eurasia (from Hungary in the west to Mongolia in the east), known as the 'Prairies' and 'Steppes', respectively. In Argentina and Uruguay, the grassland is known as the 'Pampas' and in southern Africa, the 'Veldt' (see Figure 2.2).

b) Climate

These grasslands are associated with the temperate continental climate, which has hot (or warm) humid summers, and cold (or very cold) winters, combined with low annual precipitation totals. Being more extensive, the Northern Hemisphere formations experience greater precipitation and temperature. Latitude and continentality combine to determine seasonal variations in temperature, which can attain summer highs of 38°C (although the mean is only 20°C) and winter lows of –40°C (although –20°C is more usual). Temperatures remain below freezing for 6 months every year. Precipitation is 500–900 mm/yr and drier areas are therefore semi-arid. In winter, all precipitation falls as snow and there is severe frost, freezing wind and intense wind-chill. However, most of the annual precipitation falls as rain and is concentrated in late spring and summer, which would create an effective drought in early spring if snow-melt did not ensure sufficient soil moisture during the early growth period (see Figure 5.5c).

Unpredictable rainfall, electrical storms, prolonged drought, strong, drying winds and grassland fires are all features of continental summers. These combine with high temperatures to give rise to exceptionally high rates of summertime evapotranspiration. The growing season varies from 100 to 175 days and is determined by temperature (unlike the tropical grasslands, where the determinant is precipitation). There are effectively only two seasons: a growing season when it is warm enough for plant growth and a dormant season when it is not.

c) Vegetation

Today's temperate grassland communities are composed almost entirely of grasses, herbs and flowering plants. Unlike the tropical savannas, few trees are able to colonise, except along riverbanks where soil moisture is greater. However, the present vegetation is undoubtedly a plagioclimax succession, resulting from fire and human interference across time. Where trees have successfully colonised, they are of modest height and include species such as cottonwoods, oaks and willow. The reasons for the absence of trees include:

- low precipitation, high evapotranspiration rates and the resultant negative soil moisture budget
- erratic, unpredictable rainfall
- frequent, prolonged droughts

- summer fires
- continual grazing by large herbivores which suppresses tree seedlings.

Grasses, however, are ideally adapted to thrive in such areas because:

- under drought conditions, they become dormant
- most are **perennial rhizomes**, which means that they store food and moisture in their roots and grow from buds at, or just below, the surface; such species are found in the more humid areas
- annual grasses have seeds which are drought, cold and fire resistant; they germinate quickly in spring (or after fire/drought) and are associated with areas of lower rainfall
- their narrow stems reduce heat gain and evapotranspiration
- their intricate, long root systems trap moisture and nutrients across a wide area
- their stem systems are flexible and, unlike trees or woody shrubs they can move freely in strong winds, avoiding damage to their vascular systems
- they die back in winter to avoid damage by intense wind-chill.

Such adaptations make grasses ideal colonisers of continental interiors because they allow them to:

- survive constant grazing; perennial grasses continue to grow while being nibbled
- survive fire, drought, extreme cold and strong winds as they are both xerophytic and pyrophytic.

Grasses are wind-pollinated flowering plants. Grassland ecosystems can support dozens of such species, but about 60% of the grasses within a single community will usually comprise only two or three species such as purple needlegrass, wild oats, ryegrass and buffalo grass. Flowering plants tend to be dominated by species of the sunflower and pea families. There are two different categories of grass:

- tall (**prairie**) grasses, usually attaining heights above 2 m, with root systems extending at least 1.8 m (e.g. bluestem)
- short (**steppe**) grasses, which are usually less than 60 cm high (often only 20–25 cm), but which have root systems up to 1 m deep (e.g. buffalo grass).

The grasslands of the Great Plains of North America have both types, even though the entire region is known as 'the Prairies'; taller varieties inhabit the humid east while shorter (steppe) grasses dominate in the drier west. Across the Eurasian formation, conditions are more uniformly extreme, and so the Steppes are colonised almost exclusively by the shorter grass species.

d) Soils

Grassland soils are among the most fertile in the world, equally ideal for grazing and crop production. In many areas, they have developed on deep deposits of loess, deposits of fine-grained materials such as silt and mud transported over great distances by the wind. Soil types vary across the extensive grassland areas. In drier areas, where $eT > P$ and soil moisture movement is dominantly upward, **chestnut soils** develop. In more humid areas, where $P > eT$ and downward soil moisture movement dominates, **pedalfers** such as **prairie soils** develop. Between these two extremes are the **chernozems**, the most characteristic and fertile of all the 'grassland' soils. Chernozems are the product of mull humus generated by decaying surface vegetation and root systems, which produce a characteristic black, crumbly soil (see Figure 5.5c).

During summer, decomposition is rapid and the abundant soil biota ensure that there is much mixing of the soil materials. Soil processes are arrested during winter and periods of drought. Leaching may occur in spring (during snow-melt) and as a result of intense summer storms, but this is relatively mild; any resultant downward movement of minerals is compensated for during the heat and drought of summer, when evapotranspiration causes capillary movement of soil moisture and the upward movement of minerals (including calcium as calcium carbonate) which are redeposited in the upper soil layer. Although the upper 'A' horizon is easily recognised by its dark, often black appearance, there is limited formation of lower horizons as leaching is so limited. Chernozems represent the optimum soil for agricultural use because they combine depth, fertility and excellent crumb structure. After cropping for some time, however, they may require the addition of potassium and nitrates and, if exposed, become liable to erosion by wind and water.

Pedocals such as chestnut soil develop where $eT > P$. Vegetation is consequently less dense and therefore organic input to the soil is reduced. As droughts are more frequent, there is increased capillary action and therefore soils are more alkaline. Chestnut soils can be useful agriculturally, but require irrigation and careful management; poor farming practices result in the soil quickly becoming exhausted and eroded.

In wetter areas, where $P > eT$ throughout the year and capillary action is absent, the dominant soil process is leaching. Biotic activity in such prairie soils is intense under the hot, humid conditions and, owing to soil mixing, it is difficult to distinguish between the 'A' and 'B' horizons. The indigenous tall grasses ensure sufficient input of organic matter and this is recycled rapidly during the summer months. Prairie soils are ideal for cereal cultivation.

The biomass of grasslands is low (1.6 kg/m) because of the lack of trees and woody shrubs. NPP is only 600 g/m^2/yr which is modest when compared to the savannas where grasses are taller and grow more quickly (see Figure 1.8). However, nutrient levels are high owing both to the low rainfall and to the input of the grasses' long, matted roots. Low rainfall helps to maintain the high nutrient content because leaching is minimal. Winter cold and summer drought can both arrest the recycling process, as does fire, by destroying nutrients stored within the biomass. Therefore, most nutrients are stored in either the root network or the soil (see Figure 1.9h). Because of the annual cycle of grass growth and die-back, there is a rapid turnaround of nutrients. Because there is little leaching or soil erosion, this represents a closed system unless disrupted by fire.

e) Animal life

Biodiversity is relatively low compared to the savanna grasslands, because of the intensely cold winters. However, the temperate grasslands support a high density of grazing animals, even though there is little shelter from predators. The dominant vertebrates are herbivore **ungulates** such as bison and wild horses. Such animals have hooves, which help them to run quickly when escaping predation.

In North America, the dominant herbivores are bison and pronghorn (sheep), with smaller rodents such as gophers, ground squirrels and skunks living in the safety of underground tunnels. Carnivores include coyote and badgers, larger predators such as wolves and puma having been virtually wiped out by the actions of hunters and farmers. In the Steppes, herbivores include wisent, wild horse and saiga antelope, with polecats and other members of the weasel family amongst the few surviving carnivores. Birds include eagles, geese, sparrows, owls and blackbirds, many of which migrate to escape the harsh winters. Snakes and other reptiles are common, together with large numbers of spiders, ants and other insects. Extensive agriculture has destroyed many grasslands and endangered their animal communities, many of which now require special conservation measures, for example buffalo (bison) and Siberian wild horses.

f) Human activity

Grasslands are fragile but dynamic communities. Unfortunately, the introduction of steel ploughs to cut through matted root systems has led to many temperate grasslands being cleared by their most effective predator: the farmer. The Steppes, the Prairies and the Murray-Darling Basin (in Australia) have become major cereal-producing regions, while the US mid-west and the Pampas specialise in extensive cattle ranching. The Veldt and the Canterbury plains are important sheep-rearing areas.

As grasslands vanished, their wildlife had to compete increasingly with growing numbers of domesticated livestock. Soil erosion, habitat destruction and a reduction in grazing for the indigenous herbivores have been inevitable consequences. Where overgrazing does occur, grassland may be colonised by scrub plants such as sage bush, as has happened in the drier steppes. The extermination of large carnivores often leads to population explosions among smaller herbivore species and can impact dramatically on the dynamics of the ecosystem.

Soil erosion is often the result of unsustainable cultivation techniques. For example, if dry (e.g. chestnut) soils are left bare after harvesting, the uppermost stratum becomes vulnerable to sheet erosion, especially where the relief is gentle and the winds are strong. Heavy convectional summer storms create flash floods that wash topsoil away.

Several temperate grassland ecosystems appear on World Wide Fund for Nature's (WWF's) 'Global 200 Initiative List', highlighting the dangers facing the remaining natural grassland communities around the globe. Those of particular risk include the Mongolian steppes, the North American tall grass prairies and the Patagonian grasslands in Argentina.

5 The Tundra Biome

The tundra areas are the coldest and driest of the earth's biomes; they are also the most challenging, the most fragile and the youngest. Tundra accounts for about 20% of earth's total land surface, comprising two distinct formations, one in North America and the other in Eurasia (see Figure 2.2). They are both characterised by a frost-sculpted landscapes underlain by permafrost, which restricts soil drainage. Their vegetation structures are simple and biodiversity is restricted.

Permafrost is the outstanding characteristic of the tundra biome. It is defined as any land that has been permanently frozen for longer than 2 years. As it has no fissures or pores it is impenetrable. Permanent permafrost never thaws, whereas the semi-permanent form occurs where the ground surface thaws during the summer months and is known as the **active layer**. The thickness of this layer varies according to its proximity to the North Pole, being deeper in lower latitudes.

Tundra areas are of great global significance, being one of the earth's three major carbon sinks. These are areas where the biomass absorbs more carbon dioxide than it emits. During the summer, tundra vegetation uses carbon dioxide for photosynthesis. Usually, decomposing vegetation releases carbon dioxide back into the atmosphere, but in the tundra, decomposition is arrested because of the extreme cold and so the carbon dioxide is trapped, thus reducing the amount of greenhouse gases within the atmosphere.

There are two distinct tundra biomes: **Arctic tundra** (the main focus of this section) and **alpine tundra**, which occurs at great altitude and extremely low temperatures.

a) Location

The tundra biome encircles the globe within the higher latitudes of the Northern Hemisphere, i.e. at latitudes 55–70°N (see Figure 2.2). It forms an ecotone between the Arctic ice-cap and the most northerly limits of the boreal forests covering most of Alaska, the fringes of Greenland, half of Canada and much of northern Eurasia, particularly Scandinavia and Siberia. It only occurs in the Southern Hemisphere at high altitudes, such as in the Chilean Andes. Tundra is absent from the southern polar region because the climate is too severe even for tundra vegetation to survive.

b) Climate

The most important aspect of the polar and sub-polar climates, which virtually coincide with the tundra biome, is the duration and severity of their winter conditions. Winter (when the mean temperature is <0°C) lasts for 6–10 months with mean temperatures ranging from –12 to –34°C, depending on the latitude and continentality, with Siberia experiencing the most extreme temperatures (see Figure 5.1d). Depending on the latitude, up to 2 months of winter may be continuous night, as the sun never rises above the horizon. Summers are short and mild, with mean temperatures between 3 and 12°C; this season has continuous daylight but, because the sun remains low in the sky, the Arctic experiences only low-intensity sunlight. Extremes of temperature range between –51 and +32°C, but with a mean annual temperature of –28°C. Even in summer, sudden storms can trigger a temporary return to mid-winter cold conditions and night-time temperatures plummet below freezing. The short growing season of only 50–60 days is determined as much by the thawing of the active layer as by the ambient air temperature.

Precipitation is exceptionally low, in most areas less than 125 mm, because air which is so cold can hold little or no water vapour. Tundra can, therefore, be classified as true desert. Most of the precipitation falls as snow, but rain and fog can occur in summer.

Strong, dry winds further reduce temperatures through wind-chill. However, despite such winds, low summer temperatures mean that evapotranspiration rates are also low and during the winter months, temperatures are so extreme that evaporation ceases.

c) Vegetation

Plants need sunlight and warmth to grow and reproduce, but little of either is available. Fewer plant species colonise the tundra than any other terrestrial habitat, yet, given the conditions, vegetation is relatively abundant, with over 1700 plant species having adapted successfully. In addition to the harsh climate, plants have to cope with acidic soils, permafrost and the soil movements caused by seasonal freeze–thaw. On lowland plains, plants have to adapt to wet, marshy conditions where the underlying permafrost precludes surface drainage during the summer months.

Fully developed trees are absent from the landscape (the Inuit word tundra means treeless plain), not necessarily because of the temperatures (which are often no lower than those of the adjacent boreal forests), but because of the permafrost, which cannot be breached by tree roots. Miniature tree species, such as dwarf willow, succeed because they are exceptionally shallow-rooted and less than 10 cm tall. Vegetation adaptations include:

- ground-hugging forms allowing for winter snow coverage for insulation; also, the ground surface is a little warmer than the air immediately above it
- slow growth rates
- shallow root systems because of permafrost and **solifluction**
- clustering (often in a cushion form) which helps plants to withstand low temperatures and minimise damage caused by the impact of wind-blown ice crystals
- the ability to photosynthesise in extremely low temperatures and under shallow coverings of snow
- reduction of transpiration so as to conserve water; the small leaves and thick cuticles also protect against cold, while leaf hairs trap warm air close to the leaf surface
- the albedo of Arctic plants, which allows them to absorb more solar heat than the surrounding surfaces
- plants being perennial, allowing them to store food from year to year.

Arctic and sub-Arctic plants include woody shrubs, sedges, mosses, perennial forbs and grasses, together with flowering plants such as arctic poppies, saxifrage and anemones. Lichen is also an important and widespread member of the plant community. Lichen consists of an algae and a fungus living in a symbiotic relationship making them act like a single plant; they are a pioneer species which does not need root systems to flourish on bare rock surfaces as they can absorb water and nutrients directly into their foliage. They play a crucial role in any ecosystem because they initiate soil formation, by adding organic material to degrading rock particles. In the tundra, they are also an important food source for many herbivores, including reindeer and musk ox.

The tundra biome can be likened to that of the savanna because both plant communities exhibit significant change with latitude; this is known as **latitudinal zoning**. The three recognisable 'zones' across the tundra are displayed in Figure 5.3(b).

Some tundra plants have developed mycorrhizal relationships with fungi in order to flourish in nitrogen-depleted soils. The mycorrhizae live on the roots of plants such as ling and heather, assisting the decomposition of organic and mineral matter. These nutrients pass directly into the root system, thus avoiding the soil store; in return, the fungi are provided with a relatively stable habitat among the shifting and churning soils of the tundra.

d) Soils

Tundra soils are young and immature and soil formation is an exceptionally slow process. Many writers suggest that well-developed soils are absent from the biome and certainly soil is absent from some areas such as the high Arctic tundra. There, bare rock surfaces are home to lichen and mosses with only tiny pockets of immature soil found in sheltered crannies. Across much of the middle Arctic tundra, soil accretion is inhibited by freeze–thaw action and the development of **patterned ground** (i.e. polygons of surface material formed by the seasonal expansion and contraction of water). However, within the low Arctic tundra (which forms the majority of the biome), soils have developed sufficiently to support a range of vegetation species.

In lower-lying areas, waterlogging inhibits the decomposition of organic material because of the anaerobic conditions. Consequently, layers of dead vegetation accumulate and create **peat bog**. Beneath the peaty layer is another of blue–grey clay materials, characteristic of such oxygen-starved environments where iron compounds are unoxidised. On slopes or in drier areas, podsols or gleyic podsols develop (see Figure 5.5a).

Tundra soils are characteristically waterlogged, cold, shallow and acidic. They lack clearly developed horizons, but display evidence of freeze–thaw action such as heave, creep and/or solifluction. Solifluction is responsible for the movement of shallow soils on sloping land and the development of terraces along hill slopes and valley sides; such movements inhibit soil formation. Saturation of the active layer means that plants must adapt to damp, marshy conditions. However, it provides moisture for both plants and animals, together with breeding areas for the swarms of flies and mosquitoes endemic to the area.

As a result of extreme temperatures, decomposition is exceptionally slow and so soils are low in nutrients, except where animal droppings fertilise local areas. The nutrient cycle too has adapted to environmental conditions. For up to 10 months of the year all recycling is suspended. Even during the warmer summers, chemical reactions are very slow, as are the actions of detritus eaters, bacteria, fungi and the other tiny soil organisms which drive the recycling system. Much undigested material litters the soil surface and many nutrients are 'locked' within this litter store. Few nutrients are stored within the biomass simply because the plants are so small.

The lack of nitrogen-fixing plants in the tundra limits fertility and because both chemical and physical weathering of the parent material is restricted, the input of minerals from this source is also severely limited (see Figure 1.9i).

NPP within the tundra is only 140 g/m^2/yr, making this the second lowest output after hot deserts, as can be seen in Figure 1.8.

e) Animal life

Terrestrial tundra ecosystems support only 48 resident species, but this tends to be offset by the large population of individual species. Most land animals live in a fragile yet complex interrelationship with the local vegetation and many are herbivorous in order to maximise the amount of food available to them. On land, food webs are characteristically short, but where the food webs involve joint marine–terrestrial species (e.g. the amphibious polar bear), the fifth trophic level may be reached, as shown in Figure 5.8.

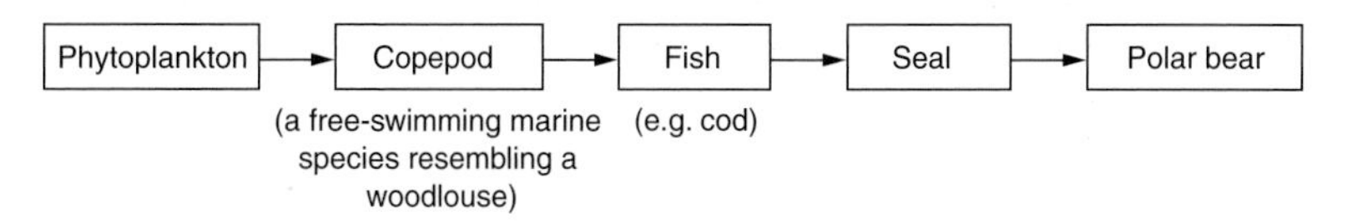

Figure 5.8 A simple food chain showing five trophic levels.

Temporary residents outnumber the permanent populations, with many bird and mammal species in-migrating during late spring. Most species are found across both tundra formations, with little localised specialisation; this is because the biome is circumpolar and migrations were uninterrupted until sea levels rose at the end of the Ice Age about 10,000 years ago.

Herbivores include lemmings, voles, reindeer (called caribou in North America), arctic hares, rabbits and ground squirrels, while the main carnivores are arctic foxes, wolves and polar bears. Insects abound and include mosquitoes, midges, moths, grasshoppers and Arctic bumble bees. Rivers are generally well stocked with salmon and trout. Most birds are summer immigrants, feasting on the abundant insect populations, although a few predators (e.g. snowy owls) are permanent residents.

Morphological adaptations to the severe conditions usually mean that animals have large, compact bodies covered by thick, insulating fur or feathers; these are often white in winter but may be brown during the summer months. For example, the musk ox (see Figure 5.9) has a large, compact body covered by a double-layered coat; the shorter hairs trap warm air to insulate the body, while the longer outer coat gives protection against wind and water. Enlarged hooves allow the oxen to break the ice cover of rivers and ponds to access drinking water during the winter.

The most common physiological adaptation is the capability to accumulate thick fat deposits during the short summer, when food is

Figure 5.9 A musk ox.

relatively abundant. These provide insulation against the cold and energy reserves for winter.

Behavioural adaptations include migration and hibernation. For example, reindeer migrate to neighbouring boreal forests during winter, returning to the tundra in spring to calve and graze the lichen cover.

Some animals such as brown bears combine two or more adaptations; during the summer, they feed heartily on the wide range of available foodstuffs, particularly migrating salmon. Consequently, thick fat deposits accumulate and provide energy throughout the bears' winter hibernation.

Marine ecosystems are much richer than their terrestrial counterparts. The Arctic Ocean has thriving, diverse populations complemented by summer in-migrants such as seabirds, which flock to the area to raise their young on the abundant marine life. Other species such as seals and walruses are permanent residents, feeding on fish and clams, respectively.

f) Human activity

Traditionally, human populations living within tundra habitats were quite small; they lived in complete harmony with the local environ-

Location	Activity
Alaska, Siberia	Oil prospecting/drilling
Kiruna, Sweden	Iron ore extraction
Alaska	Gold mining
Alaska, Lapland	Tourism
Alaska, Siberia	Transport (e.g. trans-Alaska highway and the trans-Siberian railway)
Norway, Chile	High-altitude military training (especially for US armed forces)

Figure 5.10 Economic activities in tundra ecosystems.

ment because their traditional activities of hunting, fishing and grazing large herds of caribou/reindeer were fully sustainable. The twentieth century witnessed a rapid expansion in mineral prospecting and other economic activities within these areas, with some catastrophic side-effects. Figure 5.10 displays some of the most environmentally threatening of these activities.

Expanding economic activity has increased the human population and led to construction of more towns and highways. These often block traditional migration routes, or occupy vital feeding grounds. Animals that approach settlements are frequently shot for food or out of fear. Because swarms of insects make life difficult for humans, fragile ecosystems have been subjected to the widespread use of pesticides with very little consideration of their effects on migrating birds or the food chain as a whole.

Industrialisation inevitably creates pollution. In Russia, nickel mining has wiped out expanses of wild flowers, while oil spillages threaten many habitats in areas such as Alaska and Siberia. Tank movements in Siberia as long ago as the Second World War left caterpillar tracks that are still present in the landscape. Subsequent freeze–thaw action and solifluction have deepened and widened them to such a degree that many have become deep, waterlogged gullies, which have significantly modified the local ecology.

CASE STUDY: TUNDRA NUNATAK ECOLOGY

A **nunatak** is an isolated mountain peak within a glacier or ice sheet. The nunataks in the Seward Glacier (in Yukon Territory) are inhospitable places, exposed to hurricane-force winds and temperatures forced down by wind-chill. They emerged from the ice only 12,000 years ago and their bare rock faces remain devoid of life. However, more sheltered niches support a small range of plants, animals and birds. Nunatak ecology shares many similarities with that of the tundra biome, but is even more fragile.

Plant communities include lichens, mosses, sedges, heathers and isolated alpine flower meadows, most clustered in rock hollows having southerly aspects. Poppies, daisies, dandelions and saxifrages are other common plants and attract populations of moths, dragonflies and butterflies.

Snow buntings are one of the few birds to breed on the nunataks, but about 100 species of birds including sparrows, thrushes and warblers use them as migration resting points. Many migrating birds lose their way, become exhausted or are unable to fly further because their wings become iced up. Once grounded, they cannot survive and small populations of ravens can survive by scavenging the casualties' corpses.

Pikas (small rabbit-like mammals) are the only permanent mammalian residents. They survive by sheltering among the boulders and jagged, frost-shattered rocks. Although, like rabbits, pikas are natural herbivores, they are often forced to supplement their meagre diet with dead birds, especially the brains, which provide easily digestible fat and protein.

CASE STUDY: THE SOUTHERN OCEAN BIOME

Antarctica's terrestrial ecosystems are very primitive, certainly in comparison to the Southern Ocean biome which encircles it (see Figure 1.5d). Algae, lichens and mosses can grow only where water is available during the short growing season. There are only two species of flowering plant: the Antarctic pearlwort and a grass species called Antarctic hair grass. The animal kingdom consists mainly of insects, worms, mites and single-celled life-forms called protozoa.

Of all Antarctica's inhabitants, only emperor penguins over-winter on the exposed ice sheets. Even though protected by dense plumage of feathers, which overlap like tiles on a roof, they huddle together in large groups, taking it in turns to be windbreaks against the freezing blizzards.

The below-freezing temperatures of the sea mean that many forms of marine life have additional protein in their blood, serving the same purpose as antifreeze in a car engine. Layers of fur, feathers, fat and blubber provide vital insulation and young mammals achieve rapid early growth due to their high-energy diet; conversely, adult mammals grow relatively slowly, so as minimise their energy requirements. The sea is the crucial source of food for most large predators, only using the land for resting and breeding. The top predators in the Antarctic food chain are killer whales, while krill (shrimp-like animals) are the main primary consumers within the food web. Krill thrive within the bitterly cold water of the circumpolar current that flows constantly from west to east around the continent and are especially plentiful close to the steep slopes of the Continental Shelf, where deep water rich in nutrients rises to the surface.

Antarctica possesses considerable mineral wealth and it is estimated that the Ross and Weddell Seas alone contain oil reserves similar in size to those in the British sector of the North Sea. Deposits of copper, chromium, nickel and platinum are known to exist and 3-m thick coal seams lie exposed in some of the mountainous areas. In order to protect this wilderness region from industrial development, an Antarctic Treaty implementing a range of environmental safeguards was signed as long ago as December 1959.

Antarctica's remarkable natural beauty has created a modest, but rapidly expanding tourist industry. The first small groups of tourists visited in 1958, but numbers have grown recently now average about 25,000 annually. Most travel by ship, but an Australian airline operates continental overflights. In 1991, the major tour operators agreed to enforce minimum standards to protect the fragile natural environment; the indigenous flora is so delicate that a footprint on a bed of Antarctic moss takes a full decade to fade.

Summary

- The most northerly forest biome is coniferous owing to the long, severe winters of the higher latitudes. Although few species have adapted to such adverse conditions, extensive single-species stands produce dense vegetation cover. Trees maximise their opportunities for photosynthesis by retaining foliage and adapting leaf structures to cope with extreme temperature. However, deciduous larch woodlands dominate in the coldest locations. Boreal forests display little stratification.
- Nutrient recycling in the boreal forests is extremely slow, restricted by very low temperatures and the nature of the leaf litter. The acidity of the soil is compounded by that of the humus and the underlying parent material. Leaching further intensifies acidity and creates podsols with easily identifiable horizons.
- Economic activity destroys ever-increasing areas of forest and is only partially offset by reafforestation programmes, which merely replace tree functions, not the original natural habitats.
- While the tundra biome represents a globally significant carbon sink, the coniferous forest biome does not fulfil a similarly important role because its absorption of carbon dioxide is low owing to modest growth rates.
- Temperate coastal margins are associated with deciduous vegetation; the warmer seas mitigating the cooling effect of increasing latitude.

- Like rainforests, temperate deciduous woodlands are stratified. However, it is unusual for more than three of the five possible layers to develop in a single locality. Temperate forests tend to be dominated by a single species, which is determined by local environmental factors. Although forests lack the biodiversity of their tropical counterparts, a wide range of flora and fauna inhabit such woodland ecosystems.
- Most temperate soils are deep and only slightly acidic. Some leaching does occur, but its effects are minimal because of the high clay content. Soil horizons are often indistinct because of the effective mixing activities of soil micro- and macrofauna such as earthworms.
- Widespread woodland clearance means that little true, climax vegetation remains. Even those 'ancient' woodlands that survive have been significantly affected by activities such as coppicing and grazing.
- Where maritime influence diminishes, grasslands colonise temperate continental interiors.
- Although current precipitation totals are insufficient to sustain forest growth, these grasslands represent a plagioclimax succession due to fire and anthropogenic activities rather than lack of rainfall.
- Continentality increases both annual temperature ranges and summer evapotranspiration rates; negative soil moisture budgets are common.
- Where precipitation is greater, tall grasses (**prairie grasses**) are dominant, giving rise to the popular name for the grasslands of continental North America. Where precipitation is much lower, short (steppe) grasses thrive, giving rise to the name of the Eurasian grasslands.
- Grassland soils are among the most fertile of all. Consequently wide-scale clearance has taken place to introduce extensive cereal cultivation to the wetter areas and livestock rearing to those with lower annual precipitation.
- The dominant characteristic of the tundra biome is the effect of permafrost and other periglacial land-form processes which significantly affect soil formation and colonisation.
- Climate is severe at best and very extreme elsewhere. Much-reduced insolation by the sun, intense cold and extended winters make plant succession difficult. So little precipitation occurs that these areas are classified as deserts.
- Fewer plant species successfully colonise the tundra than any other ecosystem on earth. Successful plants are typically low growing and shallow rooted; lichen is important on exposed rock surfaces. Where soils have begun to develop they are immature and often gleyed.
- Few animals make the tundra their permanent home, but large numbers of mammals and birds do in-migrate for summer.

Animals that reside permanently display a wide range of physiological, morphological and behavioural adaptations to assist their survival.

- Increasingly, human economic development is occurring in these regions, often with disastrous consequences for local plant and animal communities.

Student Activities

1. **a)** Describe *and* explain the development of forest biomes north of latitude 50°N.
   b) Give reasons for differences in the soil profiles of boreal and deciduous forests.
2. **a)** With the aid of fully annotated sketch diagrams, describe *and* explain the soil profiles associated with chernozem, podsol and brown earth soils.
   b) Using sketch diagrams, describe the recycling of nutrients in temperate grasslands, boreal and deciduous forests.
3. In what ways does climate influence the development of both soils and vegetation cover in northern temperate latitudes?
4. Describe and explain how flora and fauna have adapted to life in polar regions, basing your answer on a typical tundra ecosystem.
5. 'Vegetation succession within temperate latitudes is determined largely by temperature.' Assess the validity of this statement and justify your answer with reference to specific examples.

Small-scale Ecosystems

KEY WORDS

Bioconstruction sedimentation accelerated by the presence of vegetation
Bleaching event the expulsion of algae from coral reefs due to ecological stress
Coral reef accumulation of coral in shallow tropical seas by tiny animals (**polyps**) related to sea anemones
Dunes ridges of sand occurring in arid areas and along coastlines
Halosere plant succession evolving under saline conditions
Psammosere ecological succession that develops on sand, especially sand dunes
Salt marsh flat, sheltered coastal area that originally developed as tidal mudflats but is now vegetated
Sand particles of rock with diameters ranging from 0.06 to 2.00 mm

1 Introduction

In this, the penultimate chapter of the book, attention turns from an examination of ecosystem dynamics at the macro-(global) scale and focuses instead on smaller scale communities and habitats that develop at the micro-scale. Three particular habitats have been selected; each because of its potential importance at the global level. Each of these ecosystems plays a significant role in the protection of coastal areas from the threat of marine flooding, a threat which is growing in significance as the effects of global warming and consequent rises in sea level become ever more apparent. Yet many of these ecosystems are under threat of extinction, primarily through anthropogenic activities which fail to recognise their contribution to our continued well-being.

2 Sand Dunes: An Example of Sere Development

Sand dunes are not only effective protection against marine inundation, but also ideal locations in which to study the process of prisere development because well-established **dunes** exhibit most successional stages along a transect from the beach to their landward margins.

Dunes develop along coasts where on-shore winds regularly exceed 15 km/h and where there a continuous supply of loose **sand** that can be picked up and blown inshore (Figure 6.1). They begin to form

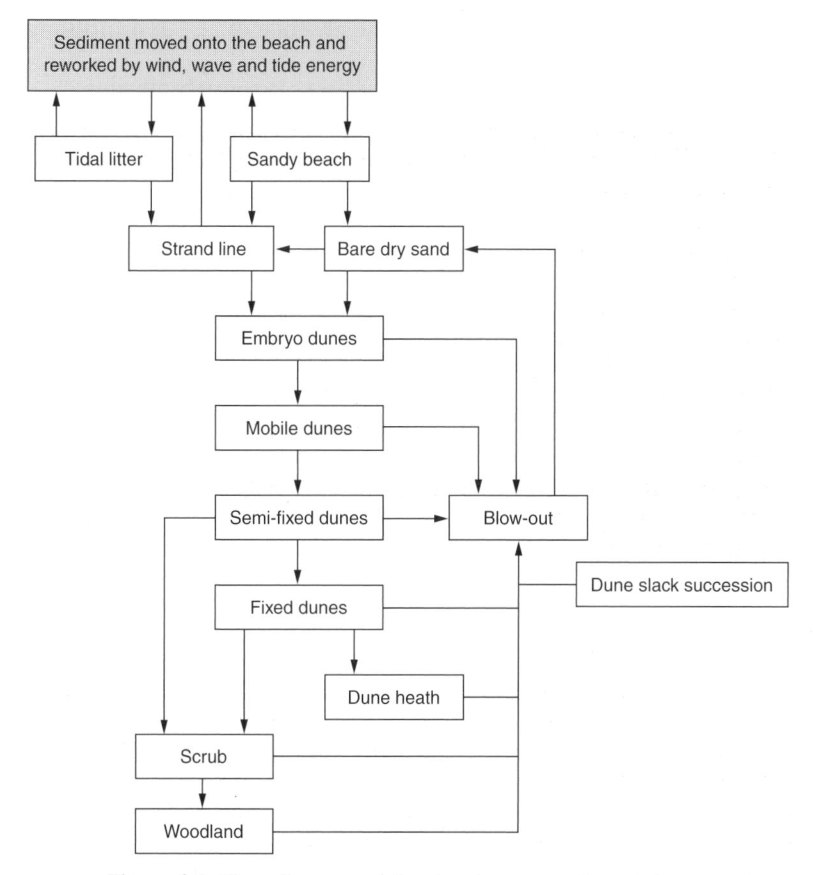

Figure 6.1 Flow diagram of the development of sand dunes.

when isolated plants or pieces of flotsam on the foreshore provide sufficient resistance to slow the wind and cause it to deposit some of its sandy load. Such deposits may accumulate to build a series of sandy ridges known as a **strand line** at the top of the beach.

Unconsolidated sand is an inhospitable medium for plant growth, partly because it remains susceptible to further wind erosion and partly because it is subject to marine inundation at high tide. It cannot provide secure root anchorage and, when transported by the wind, is abrasive to tender new foliage. Embryonic dunes are especially challenging locations for vegetation development because:

- They are unstable, and moving sand can smother pioneer plants.
- Rainfall percolates rapidly through coarse-grained sand.
- The sand is salt-laden because it has been in contact with the sea and onshore winds. This salt dissolves in rainwater, making it saline and therefore difficult for plant roots to absorb through osmosis.

- They lack plant nutrients, due to the absence of any decomposing organic matter.
- They have a high pH factor, making them extremely alkaline.

Only specialised plants can colonise such areas, species which have long, vertical roots capable of reaching sources of freshwater at the base of the sand or that have adapted to gather overnight dew. Some of these pioneer species store nitrogen in root nodules to compensate for the lack of soil nutrients.

Early pioneer species (e.g. sea rocket and prickly saltwort) are fast growing and produce networks of deep roots that anchor them in the mobile sand; they are also xerophytic, because sand dunes are arid environments even where there is plentiful rainfall. To minimise water loss and maximise water storage capacity, such plants have small, waxy leaves, which reduce transpiration, and spongy tissues that store water. All such pioneer species are annuals; this increases the survival and spread of the species because their wind-blown seeds colonise afresh each year.

If primary succession is successful and the strand line survives, embryonic dunes (fore-dunes) develop. By this time, the area is safe from inundation except during the most severe storms and therefore perennial plants are able to begin colonisation. These are usually specialised grass species (e.g. sand couch grass and lyme grass), which are highly salt-tolerant. They too have waxy leaf coatings and dense root networks. However, such grasses do not grow quickly and are likely to be buried if more than 30 cm of material is deposited over them in a year. If this does happen, marram grass will colonise the dune. While marram grass has lower saline tolerance, it is tall and flexible. Flexibility allows it to bend in high winds without suffering stem damage. Importantly, marram is able to grow through any sand deposited on it. Often, marram grass is the sole plant cover at this stage of dune development; in addition to dune stabilisation, its dense root network is critical as it initiates humus development. Humus increases the water-retaining capacity of the developing topsoil and provides essential nutrients for less xerophytic plants such as sand sedge and sea holly. Many of these species are semi-succulents, able both to store fresh water in their fleshy stems and to produce an unpalatable latex-like material to deter predation. This next stage of dune development is referred to as the 'mobile dune' or 'yellow dune' stage (see Figure 6.2).

As increasing numbers and varieties of pioneer species establish themselves across the dune area, the sand becomes more stable and fewer areas remain unvegetated. As new dunes form nearer to the beach, sand deposition ceases except under storm conditions and the marram grass dies out. Among the new colonisers is grey–green lichen that combines with the accumulating humus to give the dunes a characteristic grey patina and leads to these formations being referred to as the 'grey dune' stage of development. Other new

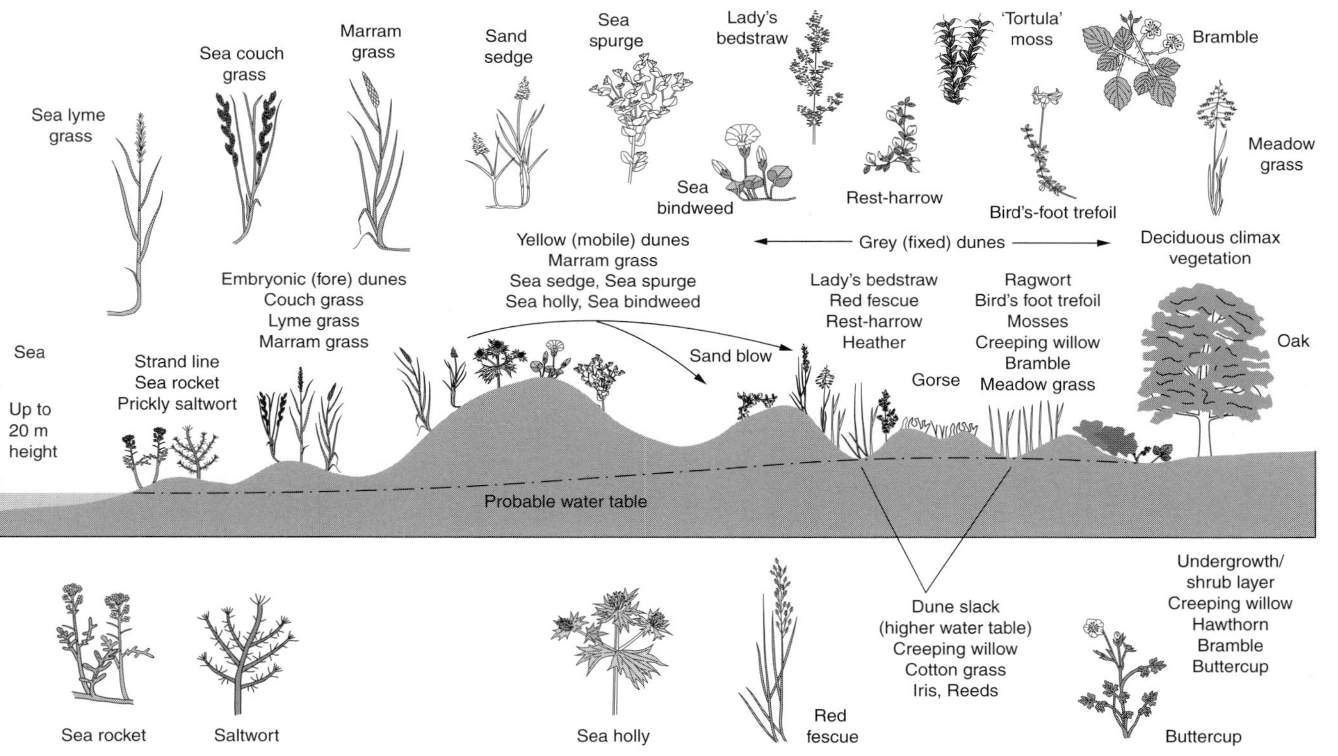

Figure 6.2 Cross-section through a typical sand dune showing the development of the vegetation succession.

colonisers include red fescue, bird's foot trefoil and rest-harrow, this latter plant being particularly important because (being a member of the legume family) it enhances the nutrient cycle; such plants have nitrogen-fixing properties and so facilitate later colonisation by woodland and grassland species.

The dunes can then develop in one of two ways:

- If the sand contains large amounts of crushed shell, there will be a higher level of bases in the soil and lime-loving plants will begin to colonise.
- If little or no calcareous material is available, the developing substrate will be acidic (because sand is pure quartz) and acid-tolerant plants such as gorse and heathers will form the next wave of succession.

Once the substrate is completely covered, dunes are regarded as being 'fixed'. Fauna such as mice, rabbits, sheep and deer may begin colonisation and attract predators such as birds of prey. If grazing is introduced, grasslands comprised of grasses, flowering plants and herbs will complete the succession because grazers crop the seedlings of any invading, higher-order, shrubs and bushes. Where grazing is limited (or where the dominant predator is removed, as happened in the 1950s when myxomatosis decimated the British rabbit population), large coarse grasses and woody plants such as creeping willow, hawthorn, bramble and sea-buckthorn will in-migrate. These species are taller and often form a canopy which shades out existing dune vegetation. Ultimately, deciduous woodland covers dunes enabling them to become an integral and permanent component of the terrestrial biome.

However, undisturbed climax woodland is extremely rare, because its development is often arrested by:

- commercial grazing
- conifer planting, aimed at reducing the amount of sand blown inland
- commercial developments such as holiday chalets and golf courses.

Dune development and colonisation is rarely achieved as smoothly as described above. One reason for this is that overgrazing leaves patches of bare sand susceptible to wind erosion; these often become enlarged into depressions called **blow-outs**, which can be so deep that they reach the water table (and are then known as **dune slacks**). Sand sedge is an ideal plant for arresting such erosion because, like other rhizomes, it reproduces and extends itself by means of subsurface runners that anchor the sand.

Although we often image sand dunes forming neat, parallel rows each at a different stage of development, this is rarely the case; more commonly they form haphazard patchworks of ridges, hollows and slacks. However, together with **salt marshes**, mangroves and **coral reefs**, sand dunes provide a cheap and highly effective means of protecting vulnerable coasts from the ravages of marine erosion during storms and high tides; their role is likely to become even more important as sea levels rise as a result of global warming.

CASE STUDY: PSAMMOSERE ECOLOGY IN AINSDALE

The Sefton coast, located between Liverpool and Southport, features one of England's most extensive dune systems (see Figure 6.3).

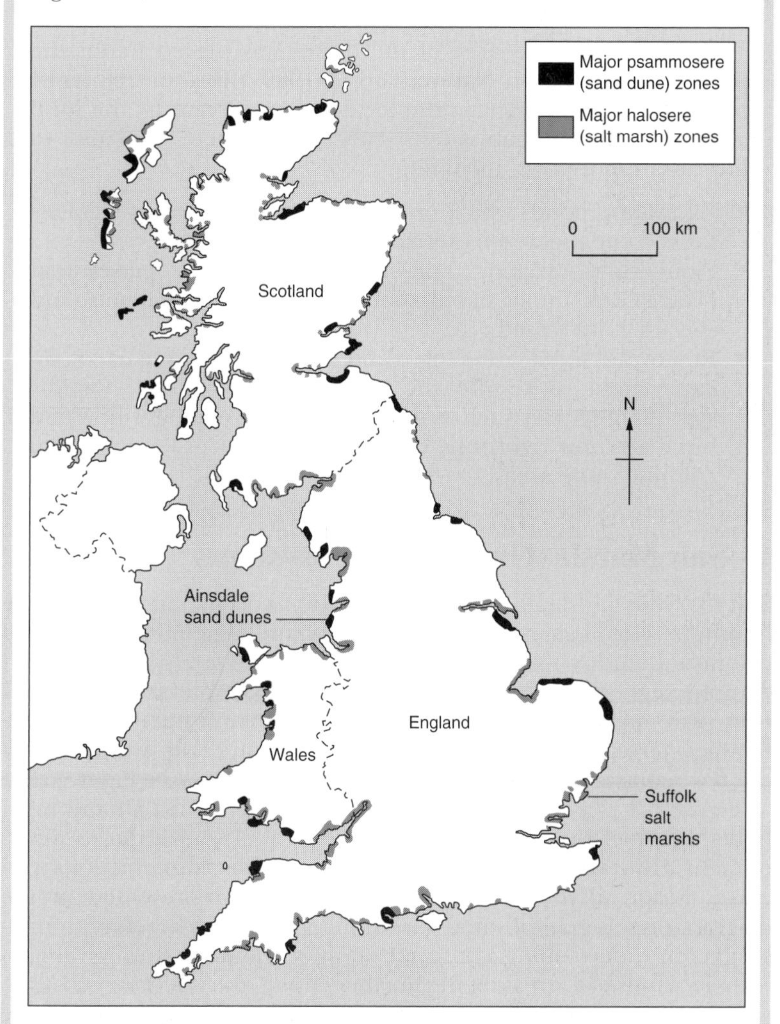

Figure 6.3 British sand dunes and salt marshes.

The Sefton coast contains separate expanses of:

- exposed beach at high tide
- sand dunes
- Corsican pine woodland.

The coastal dunes at Ainsdale provide habitats for three particularly rare, nationally protected animals: the natterjack toad, the red squirrel and the sand lizard. The more stable dunes further inland support a wide range of wildlife including foxes, rabbits, moths, butterflies and dragonflies. The woodland (which was planted at the beginning of the nineteenth century) provides a breeding ground for the endangered red squirrel.

In 1965, a 508-ha area of the dunes was placed under the protection of English Nature, and in 1980 a further 268 ha of adjacent land were declared a local nature reserve by the local authority. Both organisations have undertaken strategies to improve the reserves, including:

- Excavating ponds which provide additional breeding habitats for the endangered natterjack toad.
- Pioneering 'dynamic' conservation, which recognises that overzealous dune stabilisation can put rare habitats and species at increased risk.
- Removing some areas of planted woodland as part of an experiment to restore the original dune habitat. Rapidly spreading species such as sea buckthorn, creeping willow and birch are now rigorously controlled to allow other species to colonise dune areas.

3 Salt Marsh (Halosere) Ecosystems

Salt marshes are small, temperate ecosystems occurring mainly in the Northern Hemisphere. They develop where vegetation is able to colonise a muddy shore between high mean water (neap tide) and extreme high water (spring tide) and are an example of **bioconstruction**. Salt marsh represents an important transitional community between terrestrial vegetation and intertidal mudflats and is one of the few remaining truly 'natural' habitats in Britain (see Figure 6.3).

Figure 6.4 provides a section across a typical salt marsh and highlights the chief formation and vegetation characteristics in its development. However, salt marshes are complex webs of interactions with most, if not all, of the developmental stages represented in the diagram ongoing at different sites simultaneously. Therefore, a transect across a developing salt marsh is often a journey through time – with its landward areas being further along the successional trail, while at the seaward edge, early successional species will only just be colonising mudflats. In reality, sere development rarely continues undisturbed for any length of time, and even more rarely does it attain a mature (climax) vegetation cover. At the fragile boundary between land and sea, the destruction of embryonic salt marsh is common. Shifting river channels, variable river flow, storm surges and coastal developments are just some of the many events which arrest or destroy more advanced marshland – only for the whole process to begin again once conditions have restabilised. Today, many salt marsh habitats represent a plagioclimax due to human interventions such as grazing, reed-harvesting and engineering projects.

Salt marshes have several important functions:

- Coastal defence against marine erosion and transportation. Salt marshes and mudflats act as energy-depleting buffer-zones along vulnerable, low-lying stretches of coastline and often reduce the need for costly hard-engineering protection. For example, an 80-m wide marsh backed by a 3-m high sea wall is just as effective as a 12-m high wall without salt marsh protection in absorbing the energy of storm-force waves – representing a saving of £5000 per metre in construction costs.
- Water purification. Decomposers active within the mud strata are very efficient at removing sewage and other pollutants. By adhering to industrial and agricultural pollutants such as lead, mercury and pesticides, fine silts can prevent such chemicals from entering the food chain via marine organisms.
- Recreational provision. As more people seek refuge from the stresses of urban life, salt marshes are becoming increasingly popular places for walking, bird watching and other forms of outdoor pursuits. Creeks sheltered by mudflats and salt marsh provide safe havens for small sailing craft such as dinghies.
- Fish-breeding grounds. Salt marsh provides large quantities of nutrients for fry and molluscs (e.g. cockles and oysters). Sea bass, mullet and eels inhabit many estuaries bordering salt marsh areas.
- Nature reserves and wildlife habitats. Salt marshes boast incredible biodiversity with representatives of both the marine and terrestrial ecosystems, as well as marsh specialists. Vast numbers of indigenous birds make their homes here and thousands more in-migrate during winter. Additionally, large flocks of migrating birds use salt marsh areas to 'refuel' during long journeys. For all these reason, many salt marshes in Britain and elsewhere have been granted legal protection status, e.g. the Suffolk estuarine marshes (see page 137).

One of the greatest threats to any salt marsh is the sea itself. Major storm events can rapidly erode decades of deposition and colonisation. Other threats usually involve human action, for example:

- Land reclamation. Between the fifteenth and the eighteenth centuries, a large area of salt marsh in Britain and Western Europe was reclaimed for use as farmland. Large areas continue to be lost, but

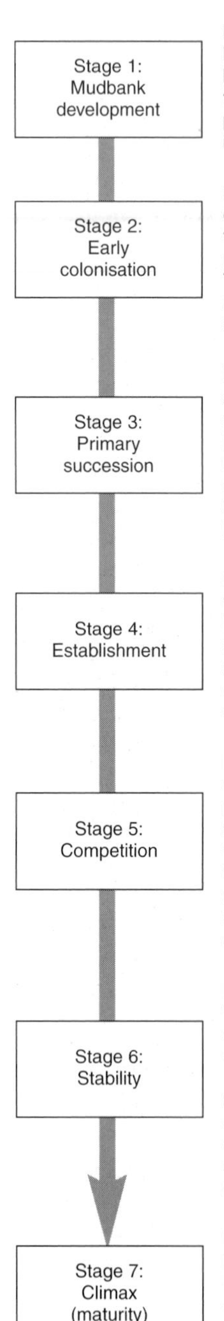

Begins along calm, sheltered shores with little wave action. Ideal locations include tidal river estuaries and areas protected by spits or bars. Fine silt (<0.05 mm) must be able to settle out of suspension; this happens with increasing salinity because sea water allows silt particles to become cohesive, enlarge and settle. As mudbanks increase in height, they become exposed twice daily at low tide.

Such **enundation** allows colonisation by halophytic algae, creating erosion-resistant surfaces which bind mud particles together. Although few species are sufficiently salt-tolerant to survive, they populate the mudbanks in huge numbers because they have so little competition. These plants trap additional layers of mud and silt, raising the mudbanks above inundating tides for 2–3 days per month, which is necessary for the next successional stage to begin.

Flowering plants begin to colonise the mudflats. Successful species are both alkaline-tolerant and halophytic because, for most of the month, tides still inundate the area twice daily. When flowering plants have gained a foothold, they alter local conditions significantly. Their roots consolidate the mud, making it more resistant to tidal erosion – and their upper fronds strap seaweed, driftwood and flotsam, thus increasing the rate of **accretion**. As flowering plants die, they add humus to the mud deposits and begin the process of soil formation.

Existing plant species grow taller and in greater numbers, thus facilitating sedimentation and extending the periods of enundation. Early successional species are short-lived but seed prolifically, thus ensuring the continuation of their own species and the whole successional process. Small herbivores visit to graze and, as developing plant communities provide increased cover from predation, small mammals begin permanent colonisation.

Early colonisers are now increasingly replaced by **equilibrium species**, which survive for longer because they have a more durable physical structure. These species have a greater chance of long-term success because although they seed later and have fewer off-spring, they produce seeds with protective casings and/or reserves of food – thus enhancing their survival rates. In the face of such competition, most of original pioneer species migrate to newly-exposed mudbanks closer to the shore-edge. Sub-habitats begin to develop, the most extensive being the level marsh (**marsh flat**).

Eventually, later successional species settle into individual niches within the habitat. Competition diminishes and fewer species in-migrate successfully. Larger herbivores such as rabbits and hares have in-migrated, together with predators such as hawks, ospreys and owls. Insects abound and attract large populations of predator songbirds such as thrushes. Waterfowl such as geese graze the herbs, grasses and flowering plants whilst wading birds visit, seeking shellfish. The saltmarsh is now rarely inundated, and only during unusually high spring tides coupled with severe storm events. Sea lavender, common at this stage, is an **indicator species** of declining salinity levels.

Bushes and trees colonise saltmarshes which reach maturity – particularly if they area have not been affected by sea-wall construction. Non-halophytic plants are able to colonise by this stage as the marshes are rarely, if ever, inundated by the sea. In reality, few marshes actually attain this level of succession! Destruction during storm events is common – and many successions are arrested by human interventions. Most saltmarshes therefore, represent a plagio-climax succession as a result of interventions such as reed-harvesting or grazing.

Figure 6.4 (a) Successional stages of salt marsh development.

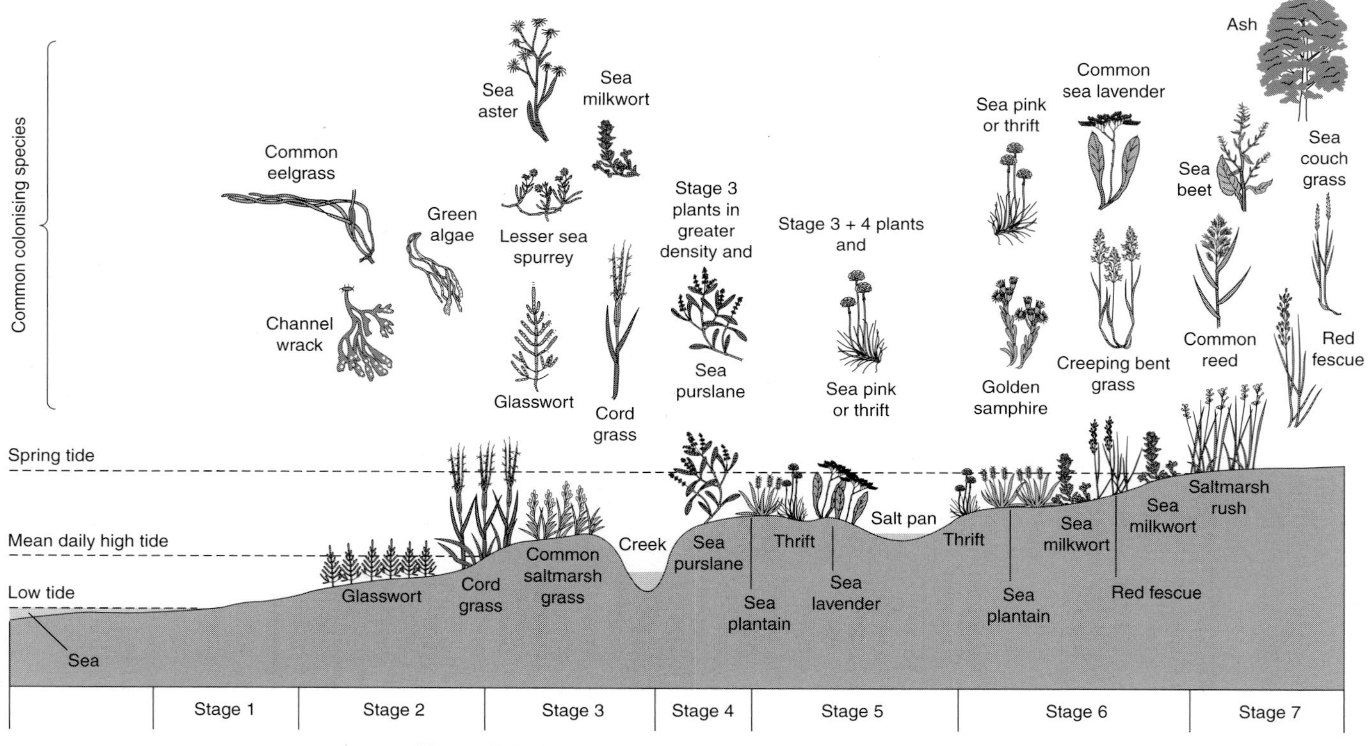

Figure 6.4 (b) Transect of a developing salt marsh.

today this is due to activities such as landfill operations, marina construction and port expansion.

- Rising sea levels. Global warming is likely to create a significant rise in sea levels, meaning that salt marshes will be drowned and protective walls behind them will be exposed to the full impact of marine erosion. When sea walls are breached, there will be extensive coastal flooding.
- Industrial development and pollution. Stabilised salt marsh provides cheap, flat land for the construction of break-of-bulk industries such as petrochemical works (e.g. ICI on Teesside). Such construction and any subsequent industrial leakages can devastate local wildlife habitats. Offshore gas and oil drilling operations pose similar environmental threats (e.g. those in the Dee Estuary and Morecambe Bay).
- Reservoir construction. Damming rivers reduces the amount of silt and clay available for estuarine salt marsh development.
- Grazing. For centuries, salt marshes have provided nutritious grazing for cattle and sheep. Overgrazing is a major reason why salt marsh often fails to reach maturity and why biodiversity is restricted. However, low-intensity grazing creates a rich grassland habitat that is highly attractive to breeding waterfowl, waders and skylarks.
- Turf production. Salt marsh provides high-quality turf that is in great demand for domestic lawns and crown-green bowling surfaces.
- Nutrient enrichment. Applying nitrogen-based fertilisers to adjacent farmland can modify salt marsh succession by facilitating invasions of exotic plant colonists.

Modern salt marsh enhancement strategies include:

- polderisation: the erection of brushwood fences to increase shelter and trap additional silt
- creating wave breakers: discarding ships and other obsolete structures offshore creates wave breaks and increases the accumulation of silt and mud
- foreshore recharging: dredged materials from harbours and shipping lanes may be deposited in locations from which longshore drift can transport it to salt marsh development areas
- the removal (or breaching) of coastal walls allows former salt marsh to revert to its former state and can reduce coastal flooding in neighbouring areas
- the (re-)establishment of salt marsh in formerly drained areas or where marsh development has been hindered; such areas may be restocked or seeded with suitable salt marsh species to foster further colonisation and marsh development.

CASE STUDY: SUFFOLK SALT MARSHES

The map in Figure 6.5 shows the chief salt marsh area associated with the five estuaries of eastern Suffolk. Four of them were formed when the sea flooded river valleys; the fifth estuary developed as the Orfordness spit grew across the mouth of the River Ore as a result of longshore drift. All five estuaries are relatively young, having been formed since the last Ice Age, and contain extensive areas of mudflats fringed by salt marsh. They are of major significance within both Britain and Europe as sites for overwintering wildfowl and waders as well as providing breeding habitats for permanent residents.

The earliest human salt marsh interventions occurred in the eleventh to fifteenth centuries, when attempts were made to reclaim land for agricultural use; a further period of reclamation took place in the sixteenth and seventeenth centuries owing to increasing population pressure.

The Suffolk salt marshes have long been a source of food and salt, as well as providing important communication and trading networks. Small-scale, early developments tended to be

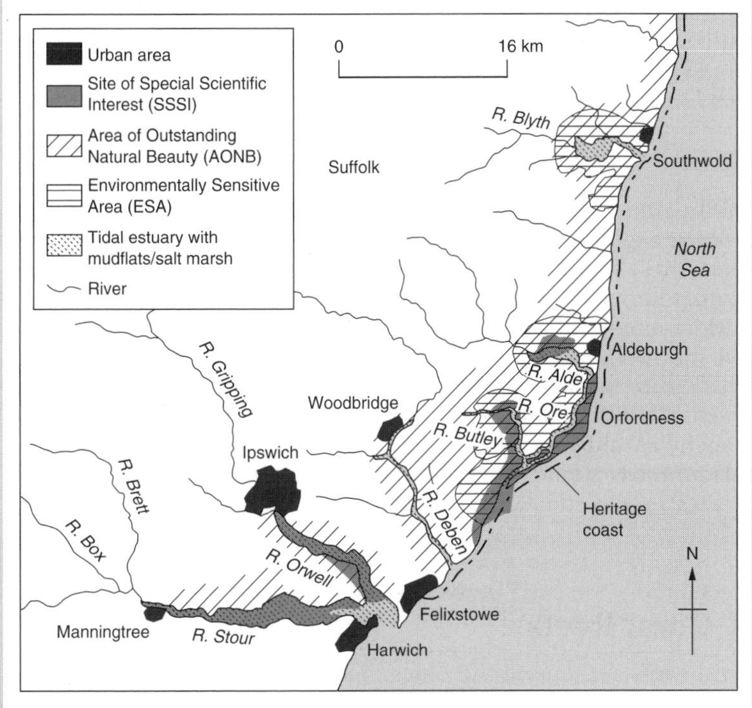

Figure 6.5 Suffolk estuarine salt marshes.

Environment threats
Port expansion at Felixstowe, Harwich and Ipswich, leading to the destruction of mud-flats and salt marsh
Dredging of navigable river channels, increasing the turbidity of estuary water, disturbing silt deposits and releasing pollutant heavy metal deposits from mudflats
Accidental, as well as some covert, illegal – discharges of oil by shipping
Leaching of highly toxic anti-fouling paints used to protect boats' hulls against marine organisms, e.g. TBT (tri-butyl tin oxide)
Discharge of industrial and domestic effluent; more than 20 sewage outfalls have polluted the River Stour alone
Discharge of fertiliser, herbicide, insecticide and pesticide residues from the surrounding agricultural land
Erosion and pollution caused by the increasing popularity of recreational pursuits on and around the estuaries, e.g. yachting, dinghy-sailing, water-skiing, wind-surfing, angling, canoeing, swimming, bird-watching, wildfowling and walking
Construction of new yacht marinas and sailing club facilities
Overgrazing of salt marsh, particularly by sheep, which results in vegetation loss and saltmarsh erosion

Protective measures
The government has banned the sale of toxic anti-fouling paints for boats up to 25 m in length and set a minimum water quality target of 0.02 micrograms of TBT per litre of river estuary water
The Stour and Orwell Estuaries have been designated Sites of Special Scientific Interest (SSSIs) – chiefly to protect overwintering migratory birds and their indigenous populations of wildfowl and wading birds
The estuaries of the Blyth, Alde, Ore and Butley have been designated part of the Suffolk Rivers
Environmentally Sensitive Area (ESA) – a measure which entitles nearby farmers to receive additional payments for undertaking traditional, environmentally sustainable farming practices
Water-skiing activities have been restricted to specified areas on the Orwell, Debden and Ore Estuaries

Figure 6.6 Threats and management issues affecting Suffolk salt marshes.

sympathetic to the local ecology but twentieth-century expansion of the region's ports has exerted greater ecological stress on the marshes. Tourist developments in the area have also had widespread impacts on local habitats, especially water-based activities, which create noise, litter and oil pollution. Water-skiing is particularly damaging as the wash increases bank erosion.

Figure 6.5 also shows those salt marsh areas which are now formally protected as a result of the initiative of conservation agencies and various European Habitats Directives; the table in Figure 6.6 summarises a range of threats to the Suffolk salt marshes and some of the initiatives taken to counter them.

4 Coral Reefs

Coral reefs are among the oldest, most diverse and most productive of earth's life zones and are the habitat of over 25% of all marine species (see Figure 6.7). They form in warm, tropical waters in areas

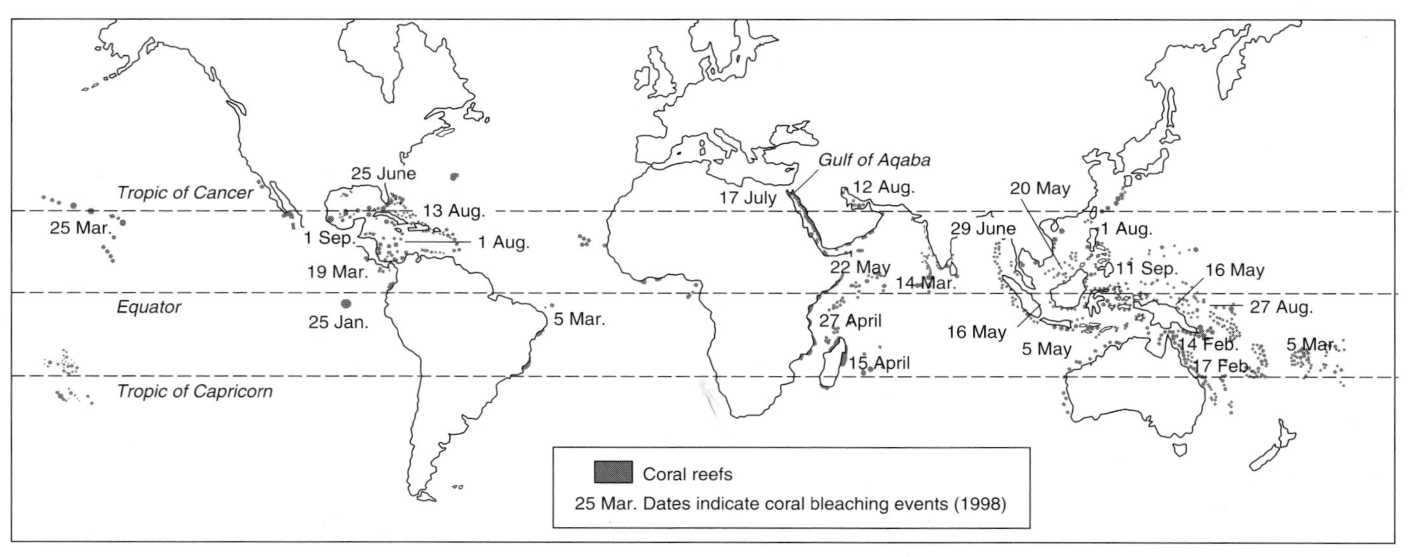

Figure 6.7 The world's main coral formations and their **bleaching events**.

of clear, shallow seas. Ideally water temperatures should be 25–29°C (although coral can survive in waters as cool as 18°C) with a salinity of 36 ppt (parts per thousand); absence of wave action and turbidity is also important. Coral is intolerant of environmental changes, particularly to variations in salinity.

Reefs are formed by massive colonies of tiny animals called **polyps** that secrete an exoskeleton of calcium carbonate around their soft bodies. When they die, these exoskeletons remain as a platform for subsequent reef growth. Coral building is a slow process and it can take over 2 years to create 2.5 cm of new structure. The maze of cracks, crevices and caves provides sheltered niches for a huge variety of marine plants and animals such as sponges, turtles, sea anemones, molluscs and fish. Some of these species use the coral for protection from predation but others, like the destructive crown-of-thorns starfish, actually feed on the coral.

Coral reefs are a joint venture between the polyps and tiny, single-celled algae called zooxanthellae, which live within the tissues of the polyp. In this mutualistic relationship, algae supplement the polyps' food intake (polyps use tentacles to catch micro-organisms floating past) and convert their carbon dioxide output into fresh oxygen through the process of photosynthesis. The polyps in return provide secure habitats and make nutrients such as nitrogen and phosphorus available to the algae which give the corals most of their bright colouring and provide food for a variety of other marine life.

Coral reefs can take several forms:

- Fringing reefs, which grow in shallow, offshore water, are separated from the coastline by a narrow stretch of sea.
- Barrier reefs, which are separated from the coast by a lagoon. They grow parallel to the coast and are large and continuous.
- Atolls, which are circular or horseshoe formations of coral around a central lagoon; atolls are usually found in deeper water.

Coral reefs provide many economic benefits, both locally and at the global scale:

- They create employment and provide food and building materials in areas where few developmental opportunities exist. This is particularly important in LEDCs, where they create employment in tourism and fishing. One-quarter of the entire LEDC fish catch is from coral reefs.
- Atolls create island homes for people in many oceanic regions.
- Their colour, beauty and biodiversity attract many tourists and they are themselves important recreational hotspots.
- They yield compounds that are effective against medical conditions such as asthma, heart disease, leukaemia and HIV.
- Like salt marshes and sand dunes, fringing and barrier reefs protect coastlines against marine erosion.

Unfortunately, coral reefs are exceptionally sensitive to any disruption in their environment and grow so slowly that, even though they can 'self-repair', this process takes a very long time. Marine biologists estimate that human activities have destroyed at least 10% of the world's coral reefs (especially in Southeast Asia and the Caribbean). Another 30% of the remaining reefs are in a critical condition and a further 30% are currently under threat. So severe are the threats to the world's coral reefs that many now have legal protection and are included in the Global 200 List of environmental-endangered areas. Regions at particular risk include the Red Sea, the Arabian Sea, the Persian Gulf, the Great Barrier Reef and the southern Caribbean Sea. Protecting reefs is difficult as well as expensive and only half of those countries with coral reefs have established nature reserves to safeguard them.

Coral reefs have always been exposed to natural disturbances such as typhoons and hurricanes, but their remarkable ability to recover from such violent, short-term threats has ensured their survival. Unfortunately, many reefs are experiencing far greater and much longer-term threats, for a wide range of reasons, many of them anthropogenic in origin:

- Sunlight may be blocked out or polyps smothered by the deposition of river-borne materials such as mining waste, eroded soil (which is greatly increased by deforestation) and offshore dredging.
- Ships' anchors, diving equipment and even bare feet weaken or fracture sections of reef. The increasing popularity of 'reef walking' is having a devastating impact on the more accessible coastal reefs, such as those described in the case study that follows.
- The increasing use of coral as a raw material in the manufacture of jewellery and cement by LEDCs.
- Pollution may be due to the increased use of pesticides and insecticides that are transported by rivers into coastal waters; the offshore dumping of raw sewage also enriches seawater, leading to increases in plankton and causing a population explosion of the crown-of-thorns starfish. Oil spillages from sea-going tankers are one of the most potent causes of coastal pollution.
- Overfishing may cause coral ecosystems to become unbalanced and allows more competitive organisms to become dominant. The practice of blast-fishing used in the Southeast Asia causes instant, widespread damage to reef surfaces; cyanide (which is toxic to polyps) is also used as a means of stunning fish.
- Increasing ultraviolet radiation from depletion of the ozone layer raises sea temperatures above reefs' tolerance levels.

Coral bleaching occurs when polyps expel the algae from their tissues as a result of stress, including higher sea temperatures and increased solar radiation. Scientists first observed this phenomenon as early as the 1920s, but there was a sudden explosion of coral bleaching events during the 1980s (see Figure 6.7) and many experts now acknowledge this as one of the indicators of global warming. Corals can recover from such trauma, but they lose their attractiveness to visitors when bleached and are then much less likely to receive the added protection accorded by nature reserve or national park status.

CASE STUDY: JORDAN'S CORAL REEFS

Aqaba is Jordan's only major settlement along that country's 32-km long coastline. Even though it is a thriving port and industrial centre, most of the population is employed in the tourist industry. The Gulf of Aqaba is a landlocked arm of the Red Sea and is the location of the world's most northerly tropical marine ecosystem. Its biggest tourist attraction is the network of coral reefs and the brightly coloured fish that live there; angel fish, seahorses, octopuses and scorpion fish are just a few of this area's indigenous species. The reefs are home to more than 100 species of coral and over 1000 species of fish, crustaceans and molluscs. This 'underwater paradise' seems permanent and indestructible, but is in fact extremely delicate and increasingly stressed by expanding tourism and other economic activities which include:

- The container port's increased trade that has wreaked havoc on the reefs' flora and led to a dramatic fall in fish numbers. Ships often have to anchor offshore, awaiting access to berths, and their anchors fracture the coral and tear long sections of it from the seabed. It is an illegal, but unfortunately common practice, for ships flush the sludge from the bottom of their fuel tanks while they are at anchor. Other pollutants include phosphates (which often spill during when loaded on ships – and are so fine that they suffocate polyps) and salts from the Dead Sea (which increase the salinity of the seawater above reefs' tolerance levels).
- Pollution from the petrochemical industries and fertiliser manufacturing that have developed close to the port.
- The laying of electricity and telecommunications cables on the seabed has cut through the coral reefs.

A number of initiatives have been undertaken in an attempt to offset some of the adverse effects of environmentally damaging developments:

- Jordan has sponsored the Gulf of Aqaba Environmental Action Plan, which encourages the development of environmental protection strategies.

- Sewage treatment plants have been constructed at Aqaba and Eilat (which is on the Israeli coast).
- Additional underwater attractions have been created to reduce pressure on the naturally formed, easily accessible reefs. In 1983, a derelict cargo ship, the *Cedar Pride*, was scuttled and is already well encrusted with coral. In 1999, obsolete army vehicles were similarly discarded and have become very popular diving targets.
- The Jordanian government has designated a 7-km, long strip of coastline as a protected marine reserve: the Gulf of Aqaba Marine Peace Park.

Summary

- Localised, coastal ecosystems such as sand dunes, salt marshes and coral reefs offer significant coastal protection from high tides and severe storm events, particularly in low-lying coastal regions. Their role is likely to become even more important with increased global warming, rising sea levels and more frequent extreme weather events.
- Sand dunes and salt marshes are ideal locations in which to observe prisere development. Both are, however, exceptionally fragile structures and decades of development can be eliminated by a single storm inundation.
- Sand dunes are an example of **psammosere** development and are colonised by xerophytic species of plants. Early colonisers also need to be halophytic, because embryonic dunes have a high salt content.
- Sand dune succession passes through five major stages: initial strand line, embryonic dunes, mobile (yellow) dunes, grey (fixed) dunes and mature (terrestrial) colonisation.
- Salt marsh succession is an example of **halosere** development. Plants colonising these habitats are halophytic as they are frequently inundated by the sea. Successful salt marsh development is heavily dependent on sedimentation. Early colonisers facilitate silt accumulation and the ensuing development is reliant on a continuing cycle of accretion and increasingly higher order succession.
- Like sand dunes, few salt marshes attain mature vegetation cover. Common human interferences include extensive grazing, hard-engineering projects and industrial pollution.
- While salt marshes – and most sand dunes – are temperate ecosystems, coral reefs are tropical, marine ecosystems.
- Coral is formed as the exoskeleton of tiny animals called polyps, which live in a symbiotic relationship with algae, and it is these algae which give the coral reefs their attractive, variegated coloration.

- Coral is harvested to meet a range of human needs and is damaged by a host of other activities, not least the tourist industry which often develops as a direct result of its inshore location.
- Polyps are exceptionally intolerant of environmental change. They quickly become stressed and then expel their resident algae, resulting in their losing their colouring and becoming 'bleached'. The final two decades of the twentieth century witnessed a rapid increase in bleaching events, regarded by many scientists as evidence of rising sea temperatures due to global warming.

Student Activities

1. a) For *either* a sand dune *or* a salt marsh ecosystem, describe and account for the development of a typical climax community.
b) In what ways might human activities arrest the vegetation succession in the ecosystem you chose for (a) above?
c) For what reasons should we prioritise the conservation of such ecosystems?

2. a) What is 'coral', and how do coral reefs form?
b) In what ways might a fringing coral reef be utilised to benefit the economy of a typical LEDC?
c) Assuming that such an LEDC has a *very low-lying* coastline, why might it be in the country's best interests to conserve its coral reefs?
d) Why do many scientists believe that coral reefs may be important indicators of global warming?

3. a) Describe *and* explain which pioneer species you would expect to find on an area of mobile sand dune?
b) Why would you expect to find greater species diversity in an area of fixed grey dunes.
c) Define the term 'dune slack' and then state which species of vegetation might be used to re-colonise such a feature.
d) Assess the likely impact of leisure and tourism on the development of sand dune ecology.

4. Describe the role of vegetation in coastal development for *either* a sand dune *or* a sand marsh habitat.

7 The Global Dimension

KEY WORDS

Acid rain precipitation that has absorbed sufficient sulphur or nitrogen to have a corrosive effect on rocks such as Carboniferous limestone
Bio-prospecting searching for potentially valuable plants for medicinal research and pharmaceutical development
Conservation the protection of natural resources and landscapes for later use
Global dimming blocking of incoming solar radiation due to atmospheric pollution
Global warming increase in the earth's overall temperatures; believed to be linked to the greenhouse effect
Greenhouse effect trapping of longwave radiation by greenhouse gases such as water vapour and carbon dioxide in the atmosphere
Pollution the existence of energy or substances that cause undesirable change in the natural environment
Principle of precaution protection measures introduced on the assumption that an environment is threatened; evidence or proof of ecological threat/damage need not exist
Sustainable development changes that maintain the ability of a population to survive over time

1 Introduction

The Millennium Ecosystem Assessment Report (published in March 2005, see www.worldchanging.com/archives/002447.html) reported that deforestation, climate change and **pollution** are compromising economic and social progress in the world's poorest nations. This is an interesting position, given that these problems are actually the legacy of economic and social progress in the world's richest nations.

2 Pollution

Pollution became an increasingly dominant aspect of life during the twentieth century in both the developed and the developing countries of the world. It is no longer solely the concern of scientists and environmentalists; most people are able to identify how pollution has an impact on their own lives. However, fewer people are able to recognise how simple changes to their lifestyle can make a significant impact on those same pollution issues.

A simple definition of 'pollution' might be: 'Any interference which disturbs environmental processes', but a more generally accepted definition is: 'An undesirable change in the physical, chemical or biological characteristics of air, water, soil or food which can adversely affect the health, survival or activities of humans or other living organisms.'

Through industrialisation, MEDCs found that pollution became part of the journey towards 'development'. Today, the threat to the continued well-being of the earth's ecosystems is increasing exponentially as more and more LEDCs join the list of industrialising nations, yet lack the wherewithal and/or political will to address the twin issues of resource depletion and pollution.

To some extent pollution is an inevitable result of human interaction with the environment. Our earliest ancestors utilised local resources and deposited their waste in the local environment; but there were few of them – and their capacity to despoil the local area was finite. Most natural habitats are robust and have the capability, across time, to heal the scars of low-grade occupation. In comparison, industrialised societies overwhelm the natural capacity of the environment to cope with their assaults on it. Originally robust ecosystems are becoming increasingly fragile, unable to adjust to the pollution inflicted by human activities and resource plundering. The transition from self-sufficiency to resource-dependent, 'industrialised' society has ignored the needs of the living environment that sustains us.

Although affluent MEDCs may have the wealth, technology and an increasing drive to move towards pollution reduction, less affluent nations are primarily driven to advance their economies; the latter may argue that they do not have the resources to justify such 'policies of conscience'. With over two-thirds of the world's population living in poverty, it is indeed difficult to see a way in which the need to conserve ecosystems can be balanced against the drive for still greater industrialisation.

3 Acid Rain and its Effects on the Scandinavian Taiga

Samples taken from ancient ice sheets show that there has always been some acidity in precipitation; this is the result of processes such as volcanic activity and the biological processes of wetland plants. The same ice masses also indicate a sudden increase in precipitation acidity, from pH 6 to 4, coinciding with the start of the Industrial Revolution (about 1760). '**Acid rain**' was first identified in 1872, during an air pollution survey in Manchester; it is defined as any form of precipitation having a pH value of less than 5. The most acidic rain ever recorded had a pH of less than 2 (slightly more acidic than vinegar).

Acid rain is caused by the burning of fossil fuels. Electricity generating stations are the chief source of sulphur dioxide (SO_2), while vehicle exhaust emissions contain the other most important source gases – the nitrogen oxides (NO_x). The potential of individual pollutants for environmental damage is greatly increased when they react with each other in the atmosphere, a process aided by direct sunlight. The resultant 'cocktail effect' gases are dissolved by droplets of water within precipitation, reaching the earth's surface in the form of acidic rain, snow, sleet or hail (see Figure 7.1).

a) Environmental effects of acid rain

The adverse environmental effects of acid rain were first observed in Scandinavia in the early 1950s, but it was not until 1967 that scientists raised public concerns about the widespread damage that resulted. Today, we are able to identify why southern Norway, southern Sweden and northern Denmark are so vulnerable to the effects of acid rain:

- south-westerly prevailing winds carrying acidic pollution from the industrial regions of Britain pass over southern Scandinavia
- because of the region's relief, most rainfall in Scandinavia is orographic
- levels of pollution in the region are increased still more by southerly winds bringing additional polluted air from the industrialised countries of mainland Europe.

The most common effects of acid rain are:

- the dissolving and leaching of nutrients and minerals from the upper soil horizons; these are essential for healthy tree growth and when absent the vegetation becomes susceptible to disease
- the destruction of the protective, waxy coating of the conifers' leaves which prevents the trees from photosynthesising efficiently
- the decline of many species of lichen because they cannot tolerate high levels of acidity; this has a knock-on effect on human activity as lichen is the staple food of reindeer; reindeer are, in turn, the principal component of the diet of nomadic communities such as the Sami of Scandinavia
- increasing levels of acidity levels in streams, rivers and lakes mean that few fish are able to survive
- birds' egg-shells become so brittle that they break under the weight of the mother during incubation.

The disappearance of any species from an ecosystem, e.g. a river or lake, affects the entire food web and reduces the survival rates of dependent species such as birds (e.g. herons) and some mammals (e.g. bears).

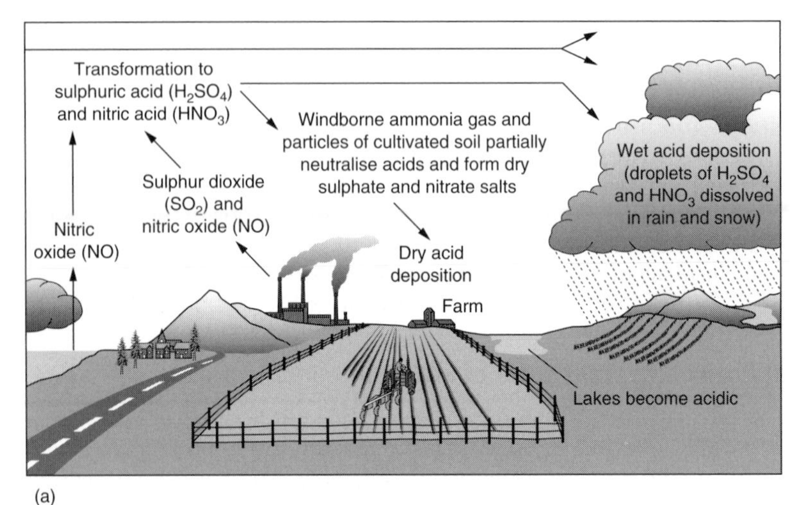

(a)

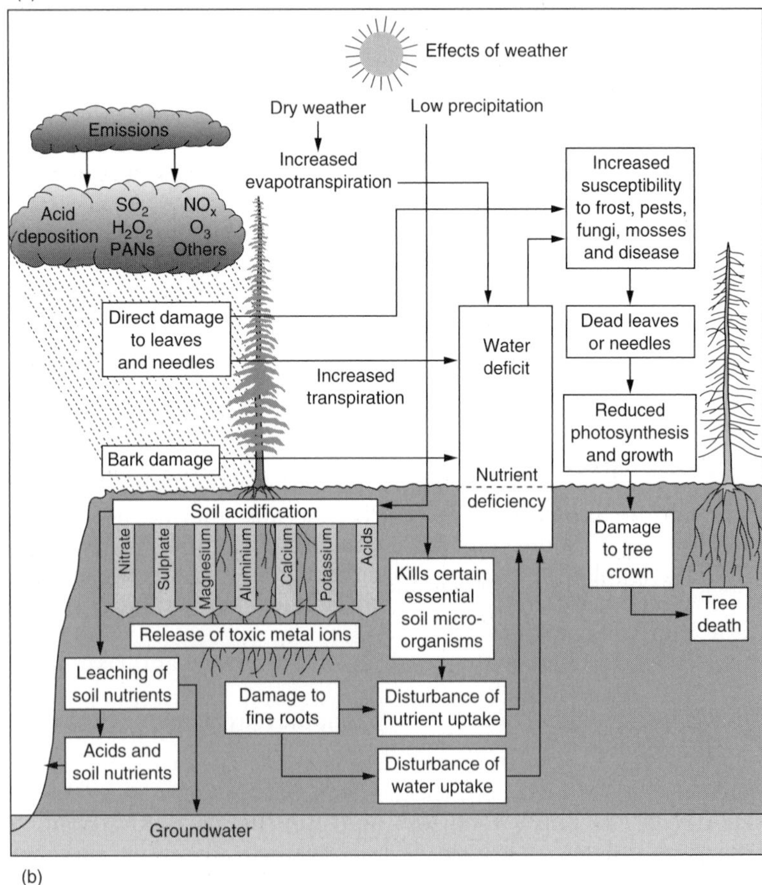

(b)

Figure 7.1 Effects of acid rain.

b) Reducing acid rain

Effective strategies for reducing levels of acid rain include:

- replacing sulphur-rich coal with other, less-pollutant fuels for industrial and domestic use. This may include using hydroelectric power, wind, wave, solar, geothermal and nuclear energy as alternatives
- fitting sulphur extraction filters to fossil-fuel burning power station to improve the quality of gaseous emissions; adding lime to react with the sulphur dioxide gas to produce solid materials that can easily be extracted
- fitting cars with catalytic converters to remove harmful exhaust gases
- encouraging travellers to use public transport instead of private cars
- adding lime to acidic lakes to neutralise excess acidity; however, although effective, it is expensive because the process has to be continued for as long as acid rain falls.

4 The Greenhouse Effect

Life on earth is sustained by the natural **greenhouse effect**, without which temperatures would be about 33°C lower, and even the oceans would be frozen. The process is represented in Figure 7.2. Current environmental concerns about the greenhouse effect focus on the evidence which appears to suggest that the process is becoming enhanced. Less long-wave radiation from the earth is escaping into space and, consequently, increased amounts are being retained within the atmosphere and reflected back to the surface.

5 The Enhanced (Augmented) Greenhouse Effect: Global Warming

Augmentation of the greenhouse effect is occurring, it is believed, because of build-up of the so-called greenhouse gases in the earth's atmosphere. Details of the causes and effects of such increases are shown in Figure 7.3. The effects of the 'minor' greenhouse gases (i.e. gases other than carbon dioxide) are believed to be the equivalent of doubling carbon dioxide concentrations within the atmosphere.

Greenhouse gases are blamed for **global warming** because increases in greenhouse gases raise the temperature of the atmosphere by trapping solar energy. However, **climate change** is a fact of life on earth; like the concentration of greenhouse gases within the atmosphere, global warming and cooling has varied across time. Changes in the earth's orbit can affect carbon dioxide levels and

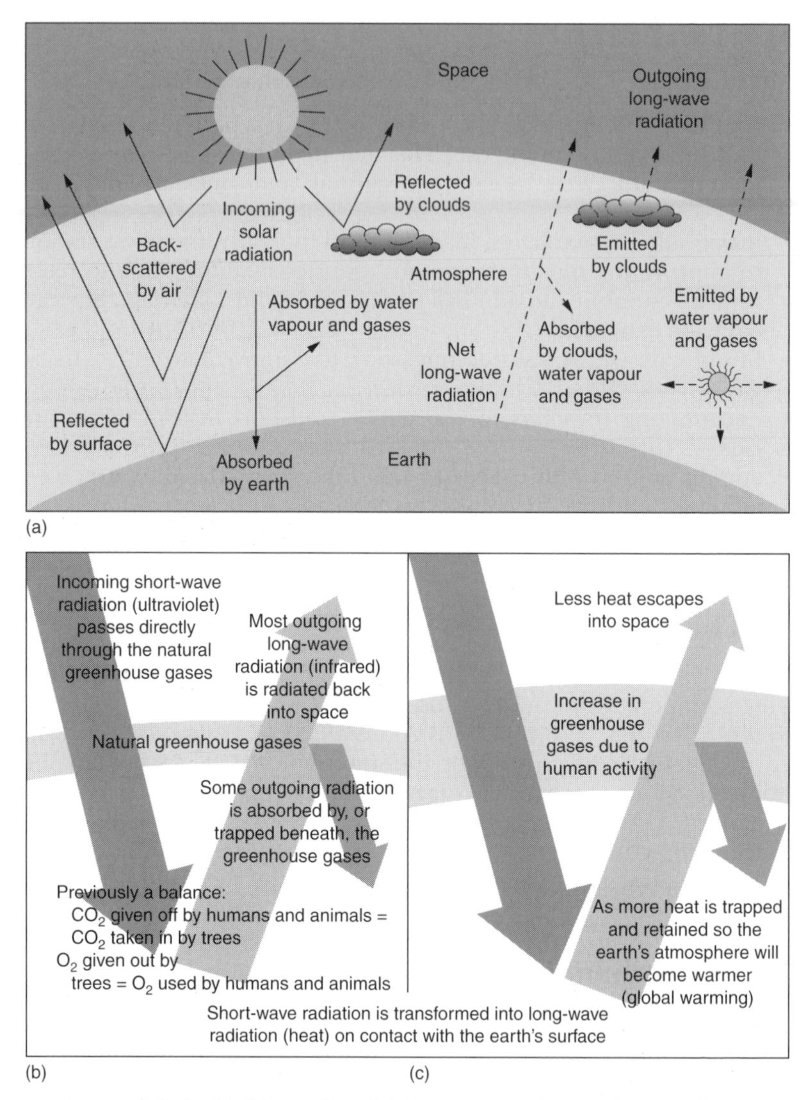

Figure 7.2 (a, b) 'Normal' and (c) 'augmented' greenhouse effects.

aerosols (i.e. small particles) in the atmosphere can also affect the climate by reflecting/absorbing solar radiation. What concerns scientists at the present time is not that the earth's climate is changing, but the unprecedented rate at which it is changing. While governments around the world may continue to dispute the evidence (or lack the political will to accept the evidence), scientists are almost universally

Greenhouse gas	Estimated proportion in 1750	Proportion in 1995	Percentage change over time	Natural and anthropogenic causes of change
Carbon dioxide	280 ppm	360 ppm	29%	Burning of fossil fuel, volcanic activity, deforestation, forest fires, organic decay
Methane	0.70 ppm	1.70 ppm	143%	Wetland processes, rice cultivation, oil and natural gas extraction, cattle and termites, landfill sites, biomass burning
Nitrous oxides	280 ppb	310 ppb	11%	Agricultural practices including use of nitrogen-rich fertiliser, burning fossil fuels, biomass burning
Chlorofluoro-carbons (CFCs)	Nil	990 ppt	Did not exist until the 1930s	Refrigerators, aerosol sprays, cleaning solvents
Ozone	Not known	Varying with latitude and temperatures	Decreasing in stratosphere, but increasing in lower atmosphere	Produced by sunlight acting on oxygen molecules and also through photochemical smogs

Figure 7.3 Greenhouse gas amounts and sources.

convinced that global warming is occurring. The evidence they use to substantiate this view includes:

- mass-melting of permafrost layers
- increasing occurrences of lethal storms and flooding, especially in South America
- increased incidences of summer heatwaves across Europe during the late twentieth century
- tree ring evidence showing increased growth
- increasing coral bleaching (see page 141)
- the rapid retreat of many glaciers
- the warming of some oceanic waters
- greenhouse gas emission data.

Attempts have been made to reduce greenhouse gas emissions at the global scale (e.g. the Kyoto Protocol and the Montreal Protocol which led to the banning of CFCs, an ozone-damaging group of chemicals), but, even if these gases are pegged at 1995 levels, this will still result in carbon dioxide concentrations twice those of 1750 levels by the end of the twenty-first century. Such a rise in greenhouse gases could be expected to:

- increase average temperatures by 2–5°C
- change global patterns of atmospheric pressure, winds, precipitation and humidity
- increase sea levels, leading to coastal flooding and increased marine erosion
- affect the polar ice-caps (which could increase in extent because of increased precipitation rather than decrease due to warming)
- increase the frequency and severity of extreme weather events
- create severe water shortages, even in the most developed countries
- shift climate zones, and consequently modify patterns of agricultural land use
- affect global processes like El Niño, exacerbating extreme weather events such as droughts and flooding.

Scientists are aware that global warming could create a massive positive feedback loop both perpetuating and accelerating the effects of warming. However, most of them admit that their understanding of the processes at work is so incomplete that they are unable to predict with certainty the extent and timing the probable global changes. Similarly, ecologists are uncertain about the consequences for plant and animal life; humans appear to be inexhaustibly adaptable, but it has taken many thousands of years for today's ecosystems to evolve. Plants are generally intolerant of change and, with accelerating global warming, it is unlikely that many of them will be able to adapt sufficiently quickly to ensure survival. Extinctions are probably inevitable. Animals may seem to be more adaptable than plants, but if their major sources of habitation and food decline, it is reasonable to assume that they too will die out in large numbers. While entire species may vanish, it is perhaps more likely that population numbers will dwindle as the more adaptable members of each species may be able to respond to changing conditions. However, ecosystems are maintained by the continuing balance of the biotic and abiotic components; if one component changes, it triggers changes in other components. Positive feedback loops could ensure that change is accelerated in directions detrimental to some ecosystems.

6 Global Dimming

Global dimming was first 'discovered' in 1985 by a geographer checking data about levels of sunlight across Europe. He realised that, compared to readings taken during the 1960s, levels of solar radiation reaching the earth's surface had dropped by 10%. At first, scientists believed that they had merely discovered an error in measuring and/or recording radiation levels but, as time passed and more comparative work was completed, it became clear that they had, in fact, discovered a *decrease* in the amount of solar radiation reaching the earth. It is now clear that solar radiation has declined by about 3% per decade since the early 1950s, not only in Europe, but throughout the world.

During the 1990s, scientists established that this decline is not the result of reduced solar output; instead, radiation is increasingly being blocked by the earth's atmosphere and is failing to reach the surface. In other words, more radiation than 'normal' is now being reflected back into space. Atmospheric pollutants appear to be the culprit, especially aerosols such as soot and sulphates, which can have two effects:

- they reflect radiation back into space
- they cause larger and longer-lasting clouds. This is a cause of concern in its own right.

If cloud formation and behaviour is affected by global dimming then there may be disruption to the world's rainfall patterns. Some scientists have gone so far as to speculate that global dimming may have been the major cause of the droughts in sub-Saharan Africa during the 1970s and 1980s. They are even hypothesising that, in the future, there may be major disruption to the Asiatic monsoon system as a result of increased industrialisation in Asia.

If air pollutants are the cause of this phenomenon, then there may be implications for agricultural yields also. Market gardeners cultivating intensively beneath glass already know that harvests are reduced by 1% for every 1% reduction in solar radiation caused by dirty glass; global dimming may have exactly the same effect on more traditional, outdoor harvests.

Of greater concern to scientists, however, is a growing worry that global dimming has been mitigating the effects of global warming. If, as seems to be the case, temperatures on earth are rising during a period of global dimming (when every hypothesis suggests they should be falling) then what might be the effects of increased global warming if solar radiation was at 'full strength'? And, more worryingly, what might be the effects on global warming in the immediate future if air pollution is significantly reduced? If air pollution were to be significantly reduced, global dimming would reduce; consequently, solar radiation would increase and global warming would accelerate as a result.

7 Desertification: The Threat of Environmental Bankruptcy

Desertification is one of the causes of great national and international debate, ranking alongside global warming, greenhouse gasses and pollution as a major threat to our continuing, comfortable habitation of the planet. Yet most people do not understand what it is. Or, they believe that it is a problem affecting small areas of the world, mainly in Africa. Most people believe that it has little impact on life in MEDCs, except to prick our consciences when we see pictures of the hungry and dying in places like Ethiopia.

Yet desertification directly affects at least 250 million people at this time. It affects two-thirds of all countries and almost one-third of the earth's total surface area. One billion people across 100 countries are at risk; that is a staggering 1 in 6 of the world's population. While the majority of those at risk are residents of LEDCs, some of the countries most affected by desertification include the USA and Spain; even Iceland has such problems. This means that, if our personal habitat is not currently threatened, our food supplies may well be. The risk is greatest in the drier areas of the world currently used for extensive cereal cultivation and stock rearing, regions which currently produce fruits and vegetables, as well as wine and other more 'exotic' produce which we take for granted.

Desertification is not (as is commonly believed) merely the encroachment of existing deserts on the lands surrounding them. Neither, it is generally agreed, is desertification solely the result of prolonged drought. Drought does have a key role to play, but it is not the causal agent. So what is desertification? A simple definition might be 'Desertification is the degradation of formerly productive land.' The UN has a more detailed definition summarised as, 'desertification is the degradation of land in arid, semi-arid and dry sub-humid areas ... caused primarily by human activity and climatic variations ...'. The key point about desertification is the significance of 'human activity'. Desertification is a 'natural' process resulting from anthropogenic activities; it is the product of the complex relationship between natural factors and human mismanagement of our most basic resource, the soil. The soil system collapses and soil erosion accelerates.

a) Natural factors in desertification

The vulnerability of land to desertification is due mainly to the interactions of:

- the climate (which affects both soil erosion and the chemical/biological deterioration of the soil)
- the topography of the land surface
- the characteristics of the soil (i.e. its texture, structure and chemical/biological properties)
- the natural vegetation. If left undisturbed, trees and bushes effectively protect against soil degradation, because they are long-lived and develop long, powerful root systems; vegetation removal

considerably increases the vulnerability of soil to erosion and thus the land to degradation.

b) Anthropogenic factors in desertification

The ways in which soil and its vegetative cover are used and managed are regarded as being the single most important causal agent of desertification; the process is primarily fuelled by changes brought about by human activity through misuse of the natural environment, for example:

- Overgrazing: better veterinary care increases herd sizes, as does the need to feed more people. Animals over-crop the available vegetation and their hooves compact the substrate. This increases fine dust particles and hence the soil's vulnerability to erosion. Additionally, herds that were nomadic are now sedentary; this trend leads to greatly increased cropping of restricted areas and further overgrazing.
- Increased demand for food: marginal land, unsuitable for agricultural activities and/or grazing is brought into use in an attempt to increase the acreage available for food production.
- Overcropping of cultivated land: the demand to increase food production in many semi-arid areas reduces the practice of fallowing and may also lead to the introduction of irrigation with its associated problems of salinisation; salinisation is yet another causal factor in the process of desertification.
- Overlogging: in the past, fuel wood was readily available from deadwood. However, increasing demand for wood (often for industrial charcoal) means that deadwood supplies are exhausted and woodland areas have to be logged to supply domestic fuel and industry. The cleared land is often left bare, precipitating soil erosion.

One of the major reasons cited for such practices is population growth; not only are there now more people to feed, but we have effectively exhausted virgin territories. In the past, when population pressure led to overusage and degradation/desertification, the populations affected simply migrated elsewhere. But now there is nowhere new to go to. Population pressures lead to unsustainable farming practices that put local ecosystems under stress and environmental degradation ensues. Agricultural yields fall and become unpredictable. The populations affected then develop 'survival strategies' to help meet their urgent need for greater food production. These increase the overexploitation of the local environment and further accelerate land degradation. This, in turn, initiates a vicious cycle of increasingly ineffective survival strategies that exacerbate the initial problem.

c) The role of drought in desertification

Drought is also a major consideration in the desertification equation, but it is not the single underlying cause. Natural vegetation in arid and semi-arid areas is drought resistant. Under normal circumstances, both natural vegetation and well-managed agricultural land will recover rapidly from prolonged drought once rains arrive. The link between drought and desertification arises in one of three circumstances:

- drought extends for an excessive period, beyond the 'norm' for that area
- the ecosystem is already stressed; environmental stress beyond the tolerance level of an ecosystem is usually anthropogenically induced
- land abuse continues during drought conditions. This is possibly the single greatest contributor to major famine events such as those in sub-Saharan Africa in the late twentieth century.

At a global level, desertification can affect us all, not only by reduced crop yields but also because of its effect on carbon exchange. Substantial amounts of carbon are currently stored in dry zones but this percentage falls when the vegetation cover is lost. As a result of carbon depletion, the greenhouse effect is more pronounced.

Desertification inevitably reduces biodiversity because it destroys habitats. Reducing biodiversity further affects the diet and the health of the local population as it reduces access to food supplies from hunting and gathering. It also destroys the gene pool of local species of plants and animals. This could be a loss for us all; many of our basic food resources were originally 'wild' plants from the drier areas of the world. With the gene pool lost, we will be unable to compensate for any in-bred weaknesses that might arise in the future. We may also lose, without knowing it, the possibilities of producing plant-based medicines to combat illnesses and epidemics. **Bio-prospecting** is still in its infancy; desertification is destroying 20,000 ha of cultivable land every year.

Desertification also affects freshwater resources. It reduces river flow and lowers underground water tables – leading to the silting of river estuaries, the infiltration of saline water into aquifers and the pollution of fresh water by salination. The biodiversity of streams, rivers, lakes and reservoirs is affected and inevitably results in declining fish harvests.

The UN believes that combating desertification is essential in order to ensure the long-term productivity of inhabited dry lands. Work in this field has been ongoing for over half a century, but most efforts have failed and the problems of land degradation have worsened. There are many reasons for these failures, some of them sociopolitical as well as technological. However, it is now increasingly

Classification of change	Descriptor
(1) Slight	Little or no degradation of soil or plant cover has occurred
(2) Moderate	(i) 26–50% of the plant community consists of climax species or (ii) 25–75% of the original topsoil has been lost or (iii) salinity has reduced the crop yields by 10–55%
(3) Severe	(i) 10–25% of the plant community consists of climax species or (ii) Erosion has removed all or practically all of the topsoil or (iii) salinity has reduced the crop yield by 50% or more
(4) Very severe	(i) less than 10% of plant community consists of climax species or (ii) land has many sand dunes or gullies or (iii) salt crusts have developed across irrigated areas

Figure 7.4 UN soil degradation classification.

recognised that effective change must start at the grass-roots level (i.e. with individual farmers) harnessing local knowledge and skills. This is a dramatic change from the former 'top–down' policies previously favoured by government bodies and international agencies. Such policies attempted to solve the problems theoretically and then 'drip-feed' strategies and technologies down to individuals and farming communities. Figure 7.4 shows the UN soil degradation classification. While the 'very severe' category in this table corresponds to most people's mental image of desertification, very few places have actually reached this extreme state. However, nearly all of the arid and semi-arid land that is currently used for agriculture does fall into Category 2 – with some areas having deteriorated into Category 3. Without early and effective remedial action, even more land will degenerate and its rehabilitation will become progressively more difficult to achieve. While scientists believe that intervention can arrest and even reverse desertification for land which falls into Categories 2 and 3, once Category 4 status is reached damage is irreversible.

8 Problem? What Problem?

Human beings are dependent on the healthy functioning of the biosphere for their continued existence. However, a vast body of evidence from many branches of science indicates that, for the first time in our history, we have developed the capability to change the dynamics of the global ecosystem and that these changes are proceeding at an alarming rate. More significant still, the majority of these changes are creating positive feedback loops, which means that they are accelerating the global system in the direction of further change instead of returning it to a state of equilibrium. These changes have been fuelled by two overriding factors:

- rapidly increasing population growth
- the demand for ever-increasing standards of living – even among the most affluent nations.

Although ecosystems have a remarkable tolerance of, and capacity for, change, ever more evidence suggests that we are pushing our global home to the absolute limits of its capacity. Scientists have been increasingly concerned about these global problems for over 30 years, and only now are politicians and the general public are awakening to the fact that:

- we are seriously damaging our global ecosystem
- if we do not change the ways in which we interact with both the biotic and the abiotic components of the environment, we will damage the earth's (and our own) life-support system.

It is not an exaggeration to state that the biosphere keeps us alive – and it does this essentially 'on the cheap'. Ecological processes shape the weather and the climate, create and generate soils and, amongst other functions, provide fresh water. Forests recycle water and act as major carbon sinks. Soils absorb toxic rain and control the release of water; at the same time, they convert minerals from the earth's crust into plant nutrients and so fuel the land-based food chains which we, as humans, head. Estuarine muds trap sediment and nutrients and host the development of important marine food chains by providing spawning grounds and nursery beds. And, without the teeming biotic communities of the soil, nitrogen cannot be 'fixed' and plant growth would diminish.

Scientists and academics at the forefront of environmental research and debate identify four critical concerns at this time:

- air pollution (because of its impact on the global climatic system)
- tropical deforestation
- the loss of biodiversity
- the increasing world population of humans (because this fuels the demands made on the earth's natural resources leading to degradation of ecosystems).

As we start the twenty-first century each day brings:

- the destruction of (or serious damage to) 300 km^2 of tropical rainforest
- the degradation towards desertification of a similar area of farmland
- the loss of 200 million tonnes of fertile topsoil
- the extinction of one species of plant or animal

- the death of 50,000 children (and a further 50,000 adults) from starvation.

It is true that humankind is adaptable; given time, we could probably devise alternatives to fulfil many of the natural processes which we appear to be destroying, but at what cost? And using which raw materials? If human activity destroys the ecosphere, there will be no second chance.

Ecological and environmental disasters such as the recurring Sahelian droughts have placed concerns about environmental processes high on the global agenda. However, some people and governments either are still not convinced about the extent of the growing problem or lack the political will to act. Former President George Bush (Sr) refused to sign the Rio Biodiversity Convention in 1992, because it was election year and US voters would not support any action that might have damaged their economy. Other national leaders refuse to acknowledge the scale of the threat using the argument that if scientists cannot give them irrefutable proof of the environmental problems, their governments are not going to change policy direction. But the scientists are as yet unable to provide such proof, mainly because they do not fully understand the complexities of interrelationships at the global ecosystem scale. There is no shortage of evidence that things are going wrong with the biosphere, very badly wrong in certain cases. But these are symptoms of the planet's lack of well-being; they are not the root causes of the problems. So this is the conundrum: scientists are struggling to find the nature of the underlying problem, to find, as the politicians demand, the proof. So far, they have been unable to establish exactly how human activities are causing the imbalances in so many natural systems. It may appear that they know enough about what's happening to take corrective action; however, if the root causes are not understood, then any 'corrective actions' that are implemented may actually worsen the situation by shifting the balance within the ecosystem in the 'wrong' direction. In just the same way that doctors have to look beyond a patient's symptoms to find the true nature of an illness before commencing treatment, scientists need to do much more than recognise the symptoms; they need genuinely to understand the processes at work.

9 Which Way Forward?

The scientists' dilemma indicates that more research obviously needs to be undertaken – and with some haste – to begin to untangle the complex workings of the biosphere. Recent technological advances in the fields of satellite imagery and computer modelling are helping and, as awareness and concern increase, funding is more readily available than it has been in the past.

With such problems to overcome, translating concern and interest into remedial action or preventive action is not proving easy. People in the economically developed world, in particular, like to consider themselves as being responsible and caring; but their treatment of the planet does not justify that belief. One change that is critical, therefore, is a change in our thinking and belief systems; without changes in our values and attitudes, as individuals, as members of local, national and international communities, effective change will not take place.

In the international arena, much is being done through the auspices of international and non-governmental agencies such as the UN the WWF and (see Figure 7.5). Large gatherings of national leaders can stimulate change; they facilitate global discussion and the exchange of ideas between scientists and national leaders in addition to attracting huge media coverage that serves to heighten awareness. They also promulgate shifts in 'management thinking'. For example, it was due to global debating forums such as Rio (1992) and Kyoto (1997) that **conservation** and global management issues are currently being addressed through concepts such as **sustainable development** and the **principle of precaution**. Sustainable development requires that any new development should meet the needs of the present while not compromising the ability of future generations to meet their own needs. This does not imply that human activity should become 'fossilised' in the present – just that its demands on limited global resources should be modest and have a long-term perspective. First advocated as a conservation principle in 1987, sustainable development was adopted as a key theme at the 1992 Rio de Janeiro Earth Summit.

Somewhat more radical is the **principle of precaution** that advocates that any proposed development involving any risk of environmental damage should not take place. If such developments are genuinely considered to be 'essential', then they should be undertaken on a small scale and the developers must build-in the maximum possible safeguards to protect the environment. The aim of this kind of 'cautious development' is to put the onus on the developers to prove that their plans will not cause harm to the environment. The concept was first published in 1992 by the UN Environment Programme (UNEP) and has since become a key aspect of many environmental strategies. It is, for example, the guiding principle in the EU's restoration programmes for the North Sea region.

One further outcome of the Rio Conference has been the requirement for all local authorities in participating countries to identify environmental problems in their communities and produce management plans that can address them, require remedial action and encourage sustainable practices in the long term. It is in response to this initiative that the conservation work now being undertaken on the Suffolk salt marshes (see Chapter 6) was authorised.

1961	**World Wildlife Fund (WWF)** founded, in Switzerland. WWF's overall aim is to reverse the degeneration of earth's environment and so help to build a future in which humankind lives in greater harmony with nature. The Duke of Edinburgh is one of WWF's longest-serving directors
1972	**Conference on the Human Environment**, in Stockholm. This was the first global conference dedicated solely to environmental matters – although boycotted by many communist countries, including the Soviet Union. Resulted in a Declaration on the Human Environment and an Action Plan to implement the Declaration
1975	**WWF's First Tropical Rainforest Campaign** Created national parks and nature reserves to converse areas of tropical rainforest in Africa, Latin America and the Far East
1980	**World Conservation Strategy (WCS)** Published by the UN Environment Programme, this document focused on purely conservation matters rather than any related political, social, cultural and economic issues
1984	**First North Sea Conference**, in Bremen – to discuss measures which may be taken to protect the North Sea's ecosystems against pollution from industrial, mineral exploitation and transportation activities. Subsequent conferences have been held in London (1987), The Hague (1990), Esbjerg (1995) and Bergen (2002).
1987	**Montreal Protocol** established restrictions for the manufacture and use of ozone-depleting substances **Bruntland Report** published by the UN Commission on Environment and Development. Its chief recommendation was that all future economic developments should be environmentally sustainable
1992	**'The Earth Summit'**, in Rio de Janeiro. 178 countries and over 1000 non-governmental organisations (NGOs) such as Greenpeace took part. Sustainable development was a key theme. The Rio Summit produced: • Conventions on Climate Change and Biological Diversity • Agenda 21 – a list of recommended ways of implementing the 27 principles agreed by delegates
1997	**The Kyoto Summit**, in Japan. Produced the Kyoto Protocol, which set a global greenhouse gas reduction target of 5% of 1990 emission levels by 2012. This was not mandatory, but was left to individual parliaments to ratify its recommendations and make them legally binding within their own countries. The USA has consistently refused to do this, stating that doing so would harm its own long-term economic growth
1991	**Millennium Ecosystem Assessment** launched by Kofi Annan, UN Secretary-General. The MEA produces reports on ecology-related topics for use by decision-makers in government and industry **WWF Global 200 List** created. Classifies 233 terrestrial, freshwater and marine realms according to their predicted levels of environmental risk
2002	**World Summit on Sustainable Development**, in Johannesburg. Undertook an audit of measures taken to increase global sustainability since the conferences held in 1972 and 1992
2005	Publication of **Ecosystems and Human Well-being: Opportunities and Challenges for Business and Industry** – the MEA's fourth in-depth report on topics of environmental concern

Figure 7.5 Steps in the right direction?

Summary

- Pollution appears to have been an inevitable result of industrialised development. The transition from self-sufficiency to industrial society has failed to take account of the fragility of the ecosystems that sustain us. Burning fossil fuels has led to acidic precipitation, which, in turn, has had a detrimental effect on both terrestrial and water-based ecosystems.
- Life is sustained by the earth's own natural greenhouse effect, but increases in the concentration of atmospheric 'greenhouse gases' have augmented this natural process and are now believed to be the cause of increasing surface temperatures. The resultant effect is known as global warming which, if continued, is likely to modify the earth's climate and biome distributions, raise sea levels and increase the occurrence of 'extreme weather events'.
- 'Global dimming' is a reduction in the proportion of solar radiation reaching the earth's surface and is believed to be a direct result of the exponential increase in atmospheric pollutants. Scientists are debating the likelihood that global dimming exerts a mitigating effect on global warming and that, as air pollution is reduced, the effects of global warming are likely to intensify.
- Desertification is degradation of land caused jointly by human activity and climatic change. Even though drought is a factor in the desertification processes, it is by no means its primary cause. Desertification is not restricted to sub-Saharan Africa, but adversely affects one-third of the earth's total land area and is therefore an issue of increasing global concern.
- Despite the increasing consensus of international opinion, some countries remain reluctant to acknowledge linkages between industrialisation and climate change. Such localised, rather than global, perspectives lead them to be overcautious about instigating measures to reduce their own countries' pollutant emissions.
- Evidence suggests that the most effective route to successful, long-term environmental improvement takes place at the grass-roots level, where local knowledge and skills can be utilised to effect sustainable change. Pollution is now a global issue – no longer a series of localised, totally independent events. An appropriate environmental mission statement for the future might, therefore, be 'Think global – act local'.

Student Activities

1. a) Explain why 'acid rain' is not confined to international boundaries.
b) Account for the potential conflicts that this can cause.

2. Describe what you understand by the following terms:
a) Greenhouse gas.
b) The augmented greenhouse effect.
c) Global warming
d) Global dimming.

3. a) Describe the views that you hold regarding the environmental issues of pollutant emission controls, global warming and desertification.
b) What actions should national governments and international organisations be taking in response to the above issues?

4. a) Define the term 'desertification'.
b) Discuss the extent to which desertification is a problem for LEDCs alone.
c) Describe the measures that can be taken to manage both desertification and its consequences.

Bibliography

All the following recommended books are published in the UK, except where otherwise stated:

Berner, E.K. and Berner, R.A. (1996) *Global Environment: Water, Air and Geochemical Cycles.* Prentice-Hall, USA.

Bridges, E.M. (1997) *World Soils.* Cambridge University Press.

Gillett, J. and Gillett, M. (2003) *The Physical Environment: A Case Study Approach to AS and A2 Geography.* Hodder & Stoughton.

Hill, M. (2002) *Arid and Semi-arid Environments.* Hodder & Stoughton.

Miller, G.T. (1998) *Living in the Environment: Principles, Connections and Solutions.* Wadsworth Publishing, USA.

Rackham, O. (1990) *Trees and Woodland in the British Landscape.* Phoenix Press.

Simmons, I.G. (1991) *Earth, Air and Water: Resources and Environment in the Late 20th Century.* Routledge, Chapman & Hall, USA.

Skinner, M. *et al.* (2003) *Complete A–Z Geography Handbook.* Hodder & Stoughton.

Trudgill, S. (2001) *The Terrestrial Biosphere.* Prentice-Hall.

Walker, J. (2002) *Environmental Ethics.* Hodder & Stoughton.

Recommended Websites

The following selection of websites provides a starting point for personal research using the World Wide Web, but it does not represent an exhaustive list of those presently available. Further sites may, of course, be obtained by initially searching very general theme topics such as 'rainforests' and 'global warming', then refining them by progressively narrowing the search field in the usual way.

www.bbc.co.uk – British Broadcasting Corporation. A useful source of topical news items and access to background information about current affairs programmes such as Horizon. For best result, enter specific details such as precise dates and locations of events/programmes.

www.independent.co.uk – The *Independent* newspaper. A useful source of topical news items often carrying environmental issues not available in other newspapers, or has a novel slant on major topics. Often gives a high profile to environmental issues and concerns.

www.royalmetsoc.org – The Royal Meteorological Society. Provides useful links to weather and climate, both in the UK and abroad.

www.woodland-trust.org.uk/main.htm – The Woodland Trust. A worthwhile source of information about British woodland issues.

www.sandsoftime.hope.ac.uk/index.htm – The Sands of Time (Liverpool Hope University College, English Nature and the Sefton Coast Partnership). An excellent site, providing detailed case-study information about the Ainsdale Sand Dunes complex on the Sefton Coast.

www.countrysideinfo.co.uk – Offwell Woodland and Wildlife Trust. Provides worthwhile information about the flora and fauna of British woodland habitats and conservation issues across a wide range of British ecosystems.

www.english-nature.org.uk – English Nature. Excellent information about UK habitats and their conservation.

www.biodiversityhotspots.org – Conservation International. Superb source of regularly updated information about environmental threats and conservation issues; particularly good with regard to fauna extinction threats.

www.panda.org – World Wide Fund For Nature (formerly called the World Wildlife Fund) (WWF). High-quality, regularly updated bulletins about global environmental issues. An excellent site from which to gather reliable and current case-study data that would be of considerable use in final examinations.

http://earthobservatory.nasa.gov/masthead.html – NASA. Offers the opportunity for interactive missions to various biomes; an interesting approach with regularly updated information. Some aspects may be simplistic for AS/A2 students, but provides good, reliable introductions to all major biomes.

www.defra.gov.uk – Department for Environment, Food and Rural Affairs. British government website covering all aspects of wildlife, countryside and rural issues.

www.rspb.org.uk – The Royal Society for the Protection of Birds. Professional site on all matters concerning wetland and other ornithological habitats.

www.geography.org.uk – The Geographical Association. A wide-ranging site covering issues of interest to both students and teachers.

www.un.org – The United Nations. Provides a global perspective on a very wide range of environmental and related issues.

Index